Project Manageme
and Techniques

CW01638265

Project Management Tools and Techniques: A Guide to Good Practice presents a unique approach to the teaching of project management by blending theory, application, and case studies. Each chapter focuses on a specific tool or technique used by the project manager in a practical and applied way. The chapters approach this by addressing the following questions:

- What does the tool or technique provide for the project?
- How is the tool/technique delivered, managed, and used?
- A case study of the tool or technique in use
- The historical context and development of the tool

This book covers core project management topics such as project feasibility, lifecycle, scheduling, budgets, risk, complexity, leadership, handover, and much more. Together the four sections of each chapter look at the benefits of using the PM process, or 'why the process is used' followed by 'how it is delivered' and then addresses 'what can go wrong if it is ignored' before noting its development as a tool within Modern Project Management. Each chapter includes a discussion of how each tool has developed using examples of landmark projects throughout the world to enable the reader to bring the theoretical and practical elements together. This new textbook covers essential skills required for successful project management, allowing both students and practitioners to apply the topics discussed directly to their environment, heightening understanding of the what, how, and why of project management. This book is suitable for use on a variety of global project management courses in disciplines across engineering, built environment, and business. It is also a useful reference for newly established PMs looking for a handy toolkit.

Aidan Turner, PhD, has worked within project and programme management for more than 30 years both in the banking, insurance, and estates sectors as well as within education. He ran Masters' level Project Management degrees at Cranfield University and currently leads undergraduate programmes within UWE's department of Architecture and the Built Environment, teaching on both these and on Masters' level programmes. His courses have been accredited by both the Association for Project Management and the Chartered Institute of Building.

Project Management Tools and Techniques

A Guide to Good Practice

Aidan Turner

Routledge
Taylor & Francis Group
LONDON AND NEW YORK

Designed cover image: Aidan Turner

First published 2026
by Routledge
4 Park Square, Milton Park, Abingdon, Oxon OX14 4RN

and by Routledge
605 Third Avenue, New York, NY 10158

Routledge is an imprint of the Taylor & Francis Group, an informa business

British Library Cataloguing-in-Publication Data
A catalogue record for this book is available from the British Library

ISBN: 978-1-032-74751-4 (hbk)
ISBN: 978-1-032-73917-5 (pbk)
ISBN: 978-1-003-47074-8 (ebk)

DOI: 10.1201/9781003470748

Typeset in Times New Roman
by codeMantra

To Ali, Josh and Joel, with love.

Contents

Introduction

This book draws on over 30 years of experience which the author has gained in both Project and Programme Management covering a variety of industry sectors and also in the delivery of related training within the higher education sector. The text presents the information in an engaging style with the aim of both helping the student to understand fully the project management process presented as the subject of each chapter and also helping the practitioner to tackle the successful delivery of the key processes and documents listed. The list of contents itself has been structured to reflect the key subjects listed by the UK's Association for Project Management (APM),[1] the USA's Project Management Institute (PMI)[2] and the Project Management Association of Japan (PMAJ) P2M[3] methodologies.

Each chapter is divided into four distinct parts. Simplistically, the aim of each chapter is to explain the 'What', 'How', 'Why' and 'Where from' questions that readers might have relating to the topic of each chapter.

The first of these four sections focuses on addressing 'what' the purpose and aim of each chapter's topic is, and it explains, drawing heavily on the three project management bodies mentioned above, what the chapter's topic means in terms of project delivery and describes the purpose of each theme. The 'how' element discusses how each topic can be managed and delivered within a project and lays out a specific approach that can be used to deliver project success, whilst also addressing some of the challenges which might be encountered.

The 'why' section in each chapter contains a different case study focused on the topic of each chapter and how the implementation of this facet of project management impacted the success of the project described in the case study. The purpose of the case studies is to draw on published reports of a project's delivery and illustrate how projects struggle if the topic of that chapter is not properly applied. It should be noted that when a project fails, there is normally more than one failing within the project management processes. The focus of the case studies in each chapter is on just one of the multiple processes which may have caused issues for the case study and is not intended to refute other potential causes of project failure. Instead, it acknowledges that the failure to apply successfully the one process in question would, on its own, have been sufficient to cause significant problems for each project case study and therefore put the project success at risk.

It is intended that the layout will enable the practicing project manager to explore the 'what', the 'how' and the 'why' of each project process and to convince themselves of the benefits of using each one. Equally, for the student studying Project Management, the book lays out in a clear and concise way the purpose, method and benefits of each of the major tools and techniques within project management and is closely focused on the kind of schedule of delivery that could be expected within a university course.

DOI: 10.1201/9781003470748-1

The last of the four sections, the 'where from' segment of each chapter, looks into the historical background of each Project Management function. It is often noted in the literature that what is now termed 'Modern Project Management' (MPM) began either in the 1950s with the US Polaris Missile Project, or with the Manhattan Project which began slightly earlier in 1942.[4,5] It was certainly after these dates that Project Management began to be codified. The APM was founded in the UK in 1972, but its first Body of Knowledge wasn't published until twenty years later in 1992. The PMI was founded in 1969 and published the first edition of its PMBOK (Project Management Body of Knowledge) in 1996. The PMAJ's first edition of P2M appeared in 2001.

The observation that the development of MPM is frequently traced back to the middle of the twentieth century gives rise to the question of how earlier, similarly complex undertakings, pre-dating either the Manhattan or Polaris projects, were managed and delivered, and how this contributed, or was a direct precursor, to the development and codification of today's MPM. This question is addressed in the fourth part of each chapter, where the findings from a case study of a large UK project delivered some 50 years before the noted beginnings of MPM are discussed.

The case study chosen is that of HMS Barham, a naval ship termed a 'Dreadnought'. HMS Barham was constructed as one of the five Queen Elizabeth Class Dreadnoughts built between 1912 and 1915. The case study was selected as it was sufficiently expensive to be considered a major endeavour at the time, with questions relating to it being raised in parliament, and additionally, it sufficiently predates the recorded start of MPM, having been built some 50 years earlier. The cost, size and national importance of the project were important parameters chosen in order to increase the chance of construction management documents having survived the passage of time and being available for research within archives.

The ship's construction meets the criteria of a modern-day project being, in the words of the APM,[1] a 'unique, transient endeavour undertaken to bring about change and to achieve planned objectives'. The APM's definition, in common with those of the PMI and PMAJ, therefore requires projects to be temporary in terms of their timescale, with a predefined start and end date which in turn implies a temporary delivery team. The uniqueness of each project inevitably adds unknowns and risks to the project which need to be managed. The planned objectives in the APM's definition relate not just to the expected outputs but also to the time, cost and performance parameters which must be adhered to.

The construction of HMS Barham faced all these challenges, thereby identifying its construction as a project that could potentially be managed in the same way as projects of today. The final part of each chapter, in reviewing the delivery of HMS Barham, discusses specific details of how the Project Management tool or technique discussed in the chapter was delivered pre-MPM as part of the construction of HMS Barham. Further background to the case study can be found in sections such as 1.4 and 2.4.

Looking beyond HMS Barham and further into history, the construction of the pyramids, Stonehenge and more recently, cathedrals in the UK, are often quoted as potential examples of early projects. They do indeed share many of the characteristics of today's modern projects, but they frequently fail to qualify as modern projects in one particular dimension. They have, nor had they, any predefined end date that they appear to have been planned to meet. Taking Salisbury Cathedral as an example, the Salisbury Cathedral website[6] states that the 'foundation stones for Salisbury Cathedral were laid on 28th April 1220', and it was 'probably completed by 1266'. Part of the reason behind the uncertainty of the completion date is that cathedrals were often continually developed over time, never seemingly reaching a 'specific objective' by a predefined end date. Turnbull[7] describes the construction of a medieval cathedral as 'the ad hoc accumulation of the work of many men', thereby lacking the structured planning required of a modern project. HMS Barham does not suffer from this issue having had a clearly defined delivery date and end state from the start, and hence is directly comparable to a project which today we would deliver using MPM methods.

Overall, the 'what', 'how', 'why' and 'from where' structure of each chapter aims to make the topic accessible and informative for the reader no matter the angle from which the reader approaches the subject. It is hoped that by providing a keen focus on the benefits that the topics discussed can bring to successful project delivery, it will help to stimulate thought and conversation for both the student and the practitioner, and enable the reader to selectively focus on the individual topics of each chapter or to follow any of the 'what'.'why' 'how' or 'from where' elements through the book.

References

1 APM. (2019). *APM body of knowledge* (7th ed.). Association for Project Management.

2 Project Management Institute. (2021). *The standard for project management and a guide to the project management body of knowledge* (7th ed.). Project Management Institute, Inc.

3 Ohara, S. (2017) P2M – A guidebook of project & program management for enterprise innovation (3rd ed.). Project Management Association of Japan (PMAJ).

4 Snyder, J.R. (1987). *Modern project management: How did we get here—where do we go?* Project Management Institute.

5 Lenfle, S., & Loch, C. (2010). Lost roots: How project management came to emphasize control over flexibility and novelty. *California Management Review, 53*(1), 32–55.

6 Salisbury Cathedral. (2021). History and conservation. https://www.salisburycathedral.org.uk/discover/history/the-cathedral-that-moved/ [Accessed 31 August 2024].

7 Turnbull, D. (1993). The ad hoc collective work of building gothic cathedrals with templates, string, and geometry. *Science, Technology, and Human Values, 18*(3), 315–340.

Chapter 1

Projects and Organisational Strategy

Over 30 years ago a report[1] focusing on the engineering construction sector recognised that projects have the potential to achieve multiple and varied strategic objectives. These strategic objectives include aims such as the regulation of internal markets, the management of subcontractors, joint ventures, the development of new areas of business and even the selling of project management expertise. Today many other commercial enterprises could be added to this list, such as the launch of new products internationally, corporate rebrandings and acquisitions and mergers to name a few.

Strategy focuses on what needs to be done, while projects focus on how the outcomes are to be achieved. Without the use of projects, competing pressures can have the potential to undermine key strategic objectives; a properly planned project, aligned to strategic objectives will maintain the focus, helping to turn chosen goals into reality.

1.1 The Importance of Organisational Strategy Within Projects

There are two types of 'strategy' discussed within the project environment. The first is encompassed in the phrase the 'strategic case for the project' while the second focuses on the strategy employed to deliver the project itself.

The APM[2] define the strategic case for the project as 'the aspirational plans, overarching purpose or intended direction of travel needed to reach an organisational vision.' The definition is focused on delivery of a desired end state for the organisation and in order to justify the project, how the project contributes to and helps to enable this end state. The PMI[3] discuss both the project's organisational strategy and the delivery strategy related to the project. The PMI list a number of 'Strategy Artefacts' detailing strategic or high-level information about the project. Key amongst these artefacts is the business case, as discussed in Chapter 2 and also the Project Brief, which provides a high-level overview of the goals, deliverables, and management processes for the project. The content of the Project Brief therefore looks at strategy from two separate angles, firstly providing a description of the desired end state, post-delivery and secondly addressing the project processes, or strategy, for delivering the project.

P2M[4] regards strategy as something which must be considered at all levels, from the organisation as a whole, through to a project's delivery strategy. A company will plan its project delivery so that it lines up, and delivers, according to the overall strategy whether it is part of a programme or a stand-alone project. P2M states that each programme or project which is subordinate to the overall company strategy or mission, also has a mission and in this way, there is

DOI: 10.1201/9781003470748-2

a hierarchy of missions, where individual projects or programmes contribute to delivering the overall corporate strategy.

1.2 How Does Organisational Strategy Impact Projects?

The PMI[3] describe feasibility as determining 'if the business case is valid and if the organisation has the capability to deliver the intended (project) outcome (benefits).' There is a stage before this which must provide a focus on how delivery of the project supports the achievement of the organisation's overall strategy. There is, therefore, a link between the organisation's strategy and the outputs from a project, both tangible and intangible, that means that once they are delivered, the project or programme outputs should help ensure that the organisational strategy is also delivered. This means that the benefits must be clearly identified and mapped directly to strategy execution. Chapter 12 addresses this in more detail.

If a project is focused on delivering benefits or outputs that support a corporate strategy, then what is a good corporate strategy? The corporate strategy should describe how an organisation can best operate in order to achieve its overall aims within a given environment. As time passes, the strategy is subject to changes both in the internal and external environments, consequently it is necessary to continually review the strategy to make sure that it, and the projects supporting it, remain relevant to the environment.

The organisational strategy is the initial means or plan by which a business seeks to maximise its long-term success. For commercial enterprises this will usually mean the maximisation of long-term profits, rather than short-term gains. For other organisations, success can be defined according to their respective purposes.

An organisation's corporate vision can focus upon many different horizons. Strategies are developed in a top-down manner and need to be understood by all those employed in the organisation. Projects are those specific activities designed to deliver these strategies.

If an organisation decides that change is needed, then it must first understand its current position – the 'as is' scenario. Once this is understood in light of the ongoing business environment, the required future state, the 'to be' state, can be defined. The act of getting from the 'as is' to the 'to be' state is accomplished through projects and the changes that they deliver.

Examples of these strategic aims are show in Figure 1.1 below. The strategic direction of the business can consist of just one strategic aim, or it could comprise of multiple directions as indicated by the intersecting circles in the diagram.

It is quite feasible that all three could be served by a project such as the vertical merger of two separate entities. By a 'vertical merger' what is meant, for example, is that company A might have a large sales force while company B might have the ability to produce at greater capacity. There is a 'vertical' relationship between the two businesses which would allow company B to increase output and for company A to sell it in new, untapped markets. In addition, by securing its ability to manufacture, company A would have a competitive advantage over its competitors. Closer co-operation and understanding of the market could also lead to new products being developed or existing ones being made more cost effectively through the development of new capabilities. Each one of these possibilities would require business process change, change that would be delivered by projects.

A corporate strategy lays out the organisation's overarching objectives and goals. It gives rise to the principal policies and plans for achieving those goals. It defines the range of business activities the company will focus on and defines the type of organisation it aims to be and the nature of its contribution to stakeholders. Figure 1.2 shows how these strategic aims breakdown

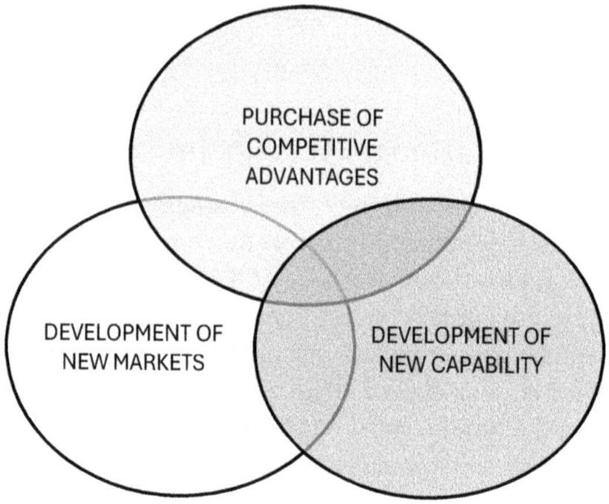

Figure 1.1 Possible strategic aims.

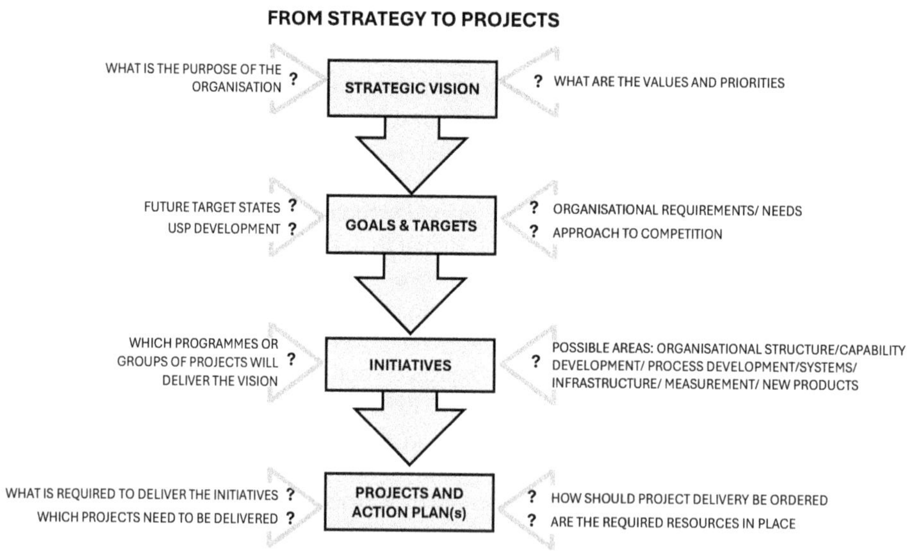

Figure 1.2 How projects derive from strategy.

into goals and targets and how projects derive from the initiatives designed to deliver the goals and targets.

The project delivery strategy for the approach to and delivery of the project, is laid out in the Project Management Plan (PMP). The PMI[3] define the content of the PMP as those processes that need to be carried out in order to complete the work necessary to satisfy the project requirements. These procedures are what are known as 'living processes' which mean that they need to be constantly reviewed, maintained and adjusted as the project is delivered in order to maintain their relevance.

The outline content of the PMP is listed below in alphabetical order, and many of these components are discussed more fully in later chapters:

- Benefit tracking and delivery.
- Budget or Cost management plan.
- Change control plan.
- Communications management plan.
- Progress monitoring.
- Project roles and organisation.
- Quality management plan.
- Requirements management plan.
- Risk management plan.
- Scope management plan.
- Stakeholder engagement plan.

If the projects are to deliver a meaningful contribution to the organisation's success, then the project delivery strategy must be aligned with the organisational strategy and must enable the project to deliver efficiently, in terms of costs, and effectively in terms of the delivery of benefits which align with and contribute to the organisational strategy.

1.3 Project Review: The Impact of Strategy on the Project – The Ford Edsel

Strategy can have many impacts on a project, whether it is the strategy of the organisation, or the strategy used to organise and deliver the project. This case study of the Ford Edsel considers a project that was, amongst other things, heavily impacted by an organisation's strategy that was fully considered, but ultimately was the wrong strategy for the times in which the project came to market.

1.3.1 The Project Background

Marketing Myopia has been described as a lack of focus on consumer requirements and an over focus on what the company itself wants to provide.[5] This results in companies going forward with a distorted view of the route to success, focusing more on what the company can produce than what consumers will buy. The mismatch produced is symptomatic of a flaw in the strategic direction of the company. This inability to understand consumer aspirations is a frequently cited cause of the failure of the Ford Edsel motor car.

In 1955, the American car market was booming. More than seven million cars were sold in the country,[6] but Ford was faced with a problem. When Henry Ford II took over control of the Ford Motor Company (FMC) in 1945 he was met with two key problems that were casting a shadow of the future of the company. Firstly, there was a lack of a modern organisational structure and system within the company, but the company also needed to address what has been identified as Ford's main goal in the early 1950s, the recovery from the long decline of the 1930s and 1940s[7] by expanding from its low-priced base and regaining its position as the largest car maker in America. It was the second of these issues that the Ford Edsel development was focused on solving. Eighty-five percent of Ford production had been in its low-priced lines in 1941, dropping to 75% by 1952.[7] Their cars were clearly selling well in the lower end of the market, but as the consumer moved towards the higher end of the market, there was a tendency for consumers to buy either Buicks or Oldsmobiles. Consumers would trade up from smaller, cheaper cars to mid-range vehicles as their income grew, which was perceived as a big issue for the Ford management who openly objected to 'growing customers' for the mid-range manufacturers.

The company was determined to adopt a strategy to address this and resolved to produce a new automobile in the 'medium-price range' ($2,400–4,000), that would compete with General Motors. To deliver on this strategic idea, the FMC spent over a quarter of a billion dollars on the production of the Ford Edsel even before the first vehicle was sold.

1.3.2 The Project Issue

Within just 2 months of the launch the car had become a laughingstock and by the time the last Edsel was produced in September 1957, the name "Edsel" had become shorthand for a massive and embarrassing public failure.[7]

Ford spent more on the launch of the new model than had been spent on that of any other consumer product in history.[6] Ford forecast sales of at least 200,000 Edsels in the first year of production alone. Two years later however, Ford had sold just over 109,000 of the new model, of which many were bought by those with a personal interest in the car's success such as Ford executives and Ford dealers. In November of 1959, after making a loss of around $350m on the Edsel, Ford cancelled production.[8]

The Edsel was designed to fit the perceived market. It was purported to have been designed as a result of much consumer research, but evidence shows that it was instead designed in line with the hunches of company committees. Similarly, the Edsel name was supposed to be derived from considerable market polling as was the approach and direction of the marketing and promotion of the car. These however were also both based upon intuitive methods rather than drawing upon public market research.[6]

The car had been fully styled by the middle of 1955, it had a novel, horse-collar-shaped radiator grille, set vertically in the centre of a conventionally low, wide front end,[8] design aspects that were intended to be recognised as a blend of the familiar with the unique. The Edsel had other unique features such as the appearance of the car from the rear and the cluster of automatic transmission buttons on the steering wheel, all of which added to the space age feel. Two years after the styling was completed, in September 1957 the Edsel was finally made available to the public.

It was, at least in part, the two years spent between design and getting to market that proved to be the Edsel's undoing. In these two years the taste of the consumer had changed. Smaller and less powerful cars, known as 'compacts', had become popular, popular enough to mean that the concept of a large car such as the Edsel did not suit the 1957 market.[8] This mismatch is perhaps even easier to understand given that while there were 2 years between the completion of the styling and the first sale, there were 7 additional years between the original studies to determine if a new and different medium-priced car should be developed and the completion of the first design in 1955.[6] Since these first studies were conducted, the Korean War which began in 1950 meant that Ford's plans for the car were put on hold for 2 years. Approximately six months before the end of the war, in late 1952 studies were resumed and in December of 1954 Ford's Forward Product Planning Committee predicted that by 1965 America's Gross National Product would increase by a third to $535 billion in the ten years from 1955 (GNP actually increased by 70% in the 10 years to 1965). The resulting increase predicted in the number of cars that would be sold was 20 million, and increasing salaries would mean that over 40% of the cars sold would be in the medium-price range.[6] If Ford didn't bring a successful medium-priced car to market, it was 'clear' that the company would miss a big opportunity.

The 1957 launch of the Edsel therefore went ahead and was part of a company restructure which saw Ford's other brands, the Lincoln and Mercury divisions, given more autonomy as part of a plan to decentralise operations.

Delays and quality problems however plagued the launch of the new Mercury and Lincoln models and public reaction to the new models was poor. Sales fell and dealer profits were impacted, dropping by more than two-thirds in 1956, resulting in almost a third of dealers making a loss.[7] Four days before the Edsel was introduced, Mercury and Lincoln were collapsed back into one single division and Jack Reith, one of the main proponents of the plan, was demoted and consequently left the company. This called into question Ford's strategy at a corporate level, which had relevance for Edsel, it had also been set up as a separate division, yet the new divisions for Ford's existing marques were no longer being supported.

At a project level, the strategy for the Edsel was also called into question. The low sales reported above, and public response to the new model suggested that the car might not be the market share saving success that Ford had hoped for, and that had been predicted by Ford's Forward Product Planning Committee. The early discontinuation of production in 1959 confirmed this. Ironically the time spent setting up production and designing the new car was greater than the time that the car was on sale to the public! The poor reception for the new model was partly caused by the defects, both real and perceived, of the car. Compared to its competitors it was overpriced, perhaps due to the large investment of a quarter of a billion dollars, spent before the car came to market. It was also perceived as having lower reliability than its competitors, perhaps due to the gadgetry and innovations included in the new model. Its distinctive styling, thought by the designers to be one of the main selling features, was largely rejected by the consumers who considered the styling odd rather than the image of elegance, power, and confidence that the designers envisaged.[7] One commentator went as far as to say that the Edsel reflected the aims and feelings that the automobile industry "have been trying to foist upon the majority of the public, fabulously overpriced jukeboxes such as only people of depraved origins or the neurotic would want to buy and only the prosperous can afford to maintain."[9]

The aim of the project and the degree to which the strategy was able to deliver against Ford's overarching goal of developing market share was flawed. The end product was outdated, costly to bring to market, and did not reflect consumer aspirations once it was publicly available.

1.3.3 How Could an Effective Strategy Have Helped?

There were two main issues with the strategy used to develop the Edsel. Firstly, the time taken to deliver the project from the idea's original inception in 1948[6] and secondly the lack of engagement with the public in terms of the focus groups.

The product development fell victim to the hubris of company executives. It was the wrong design at the wrong time. Having been designed in line with the aspirations of the company executives and committees, more investment in consumer research would have updated views on the original studies conducted in 1948, the final design in 1955 and entry to the market in 1957. In short, a more accurate, current strategy would have meant that a product could have been launched which more accurately met the aspirations of the American consumers. The time lag meant that the market was misjudged, indeed only a few years after the launch of the Edsel, the smaller, less powerful 'compacts', became so popular that the automotive 'status ladder' was reversed.[6] Amazingly, Ford's Executive Committee began planning the withdrawal of the

Edsel's from the market as early as April 1958, just seven months after its launch.[7] Time magazine went as far as to write that 'the Edsel was a classic case of the wrong car for the wrong market at the wrong time',[10] which is a very clear inditement of the project strategy.

1.3.4 Conclusions on the Edsel Project

There were numerous causes of the car's failure. The car was heavily marketed but one Edsel division executive, once they had left the company, estimated that only about 50% of the first Edsels really worked properly.[10] This was inevitably a source of negative publicity for the car. If this could have been addressed, and if the car had come to market a few years earlier when the economy was really booming and there was the demand for this type of medium-priced car, then it might well have been a success.

Ultimately, though Ford's failed attempt to successfully introduce a new medium-price range car was due to its lack of an in-depth analysis of the environmental variables governing its market. By the time the Edsel was available its appeal as a concept was waning, the public requirement had changed and they were beginning to want more simple, compact cars,[6] as illustrated 'across the pond' by the 1959 launch of the hugely successful Mini.

The failed Edsel project is estimated to have lost Ford around $350 million, impacting the market price of Ford's stock in 1958, which fell by a third, from a 1957 high, reflecting a decrease in net income per share of approximately 60%.

1.4 Organisational Strategy and Projects Before the Birth of Modern Project Management*

(* In this section, the research relating to the construction of HMS Barham 1912–1915 mentioned in the introduction is drawn upon to help illustrate how the subject of this chapter has developed.)

Before the Birth of Modern Project Management (MPM) many large undertakings were, of course, successfully delivered as discussed in the introduction to this book. Not all were projects, but the construction of HMS Barham, built between 1912 and 1915, does meet the criteria, discussed in the introduction, of a modern project.

The construction of HMS Barham was part of the bigger aim of maintaining a strong navy during both the lead up to and throughout during the First World War.

The perceived importance of maintaining British naval dominance was reflected in the 'Two-Power Standard' that determined the acceptable size of the British Fleet. This standard was defined by Earl Cawdor in the House of Lords, quoting the Prime Minister of the day, as being 'a preponderance of 10 per cent over the combined strengths in capital ships of the two next strongest Powers'.[11]

Documents revealing a specific project strategy relating to HMS Barham, have not been found during the archival research, but elements of a strategy can be seen. Figure 4.11 describing key events in the construction of HMS Barham shows the sequence of these events and the presentation of the Naval Estimates, detailing the cost of each year's programme of shipbuilding, to parliament each year, shows the approval of the strategic approach.

The strategy used to deliver the Dreadnoughts (of which HMS Barham was one) was to generally build four or five similar ships each year, the funding of which, termed the 'Navy Estimates' would be debated and generally approved, in the UK parliament. Each year during this period, a particular 'Class' or type of ship would be ordered. In 1912, it was the turn of the Queen Elizabeth Class ships. HMS Barham was the fourth of the Queen Elizabeth Class (HMS Queen Elizabeth, HMS Warspite, HMS Valient, HMS Barham and HMS Malaya.) The first two were laid down in 1912 in HM Dockyards Portsmouth and Devonport respectively. The remaining three were laid down in commercial yards in 1913, these being the yards of Fairfield, John Brown and Armstrong Whitworth respectively. The three were constructed in private yards meaning that the target of four or five ships per programme could be met. Construction was under the eye of the Admiralty Inspector to ensure certain standards were met as the construction developed, and to a degree, that lessons learnt at the HM Dockyards could be transferred to the commercially built ships.

Clearly HMS Barham was a small part of the overall First World War strategy. None the less, a successful part of the strategy was the approach and manner in which the construction of HMS Barham and the five vessels in the Queen Elizabeth class as a whole was delivered.

References

1 Lord, M.A., 1993. Implementing strategy through project management. *Long Range Planning*, *26*(1), 76–85.
2 APM. (2019). *APM body of knowledge* (7th ed.). Association for Project Management.
3 Project Management Institute. (2021) *The standard for project management and a guide to the project management body of knowledge* (7th ed.). Project Management Institute, Inc.
4 Ohara, S. (2017). *P2M – A guidebook of project & program management for enterprise innovation* (3rd ed.). Project Management Association of Japan (PMAJ).
5 Sharma, M. (2015). Marketing myopia–A mousetrap that marketer should avoid. *International Research journal of Management Sociology Humanities*, *6*(4), 488–491.
6 McCampbell, A.S. (1997). Strategy for new product development: How to avoid failures. *Journal of Customer Service in Marketing & Management*, *3*(1), 39–58.
7 Dicke, T. (2010). The Edsel: Forty years as a symbol of failure. *Journal of Popular Culture*, *43*(3), 486–502.
8 Brooks, J. (2014). *Business adventures: Twelve classic tales from the world of wall street*. Open Road Media.
9 Hayakawa, S.I. (1958). Why the Edsel laid an egg: Motivational research vs. the reality principle. *ETC: A Review of General Semantics*, *15*, 217–221.
10 Warnock, C.G. (1980). *The Edsel affair—What went wrong*. Pro West.
11 *Hansard*. (1908, November). HL Deb vol 197 cols 25–65.

Chapter 2

The Business Case and Investment Appraisal

This chapter focuses on one of the most important project documents. Once agreed, it becomes a 'living' document having the potential to evolve throughout the project's lifecycle. It should be closely monitored by the project manager and must reflect any changes which are agreed upon in terms of project deliverables.

2.1 The Purpose of the Business Case

The business case, in its initial form, is one of the first project management documents to be drafted. It is key to providing a firm base for the project in numerous ways. The Association for Project Management[1] state that the business case provides the justification for any Project or Programme. It does this by weighing up the benefits which are sought by the client and balances these against the costs and risks involved in delivering these benefits. It should assess the different options and the the different pathways, to delivering these benefits and provide a reasoned approach as to why the chosen project solution has been selected.

The benefits, the APM tell us, would normally have a tangible value, and so can be balanced against the monetary costs of project delivery. The phrasing of this definition would suggest though, that there are occasions where there can be no tangible or monetary value associated with project benefits. It is probably nearer the truth that benefits become harder to monetise. Consider for example a project that is commissioned to ensure that an organisation can meet the requirements of new legislation. Without this compliance the organisation will no longer be able to trade. The benefits here are hard to measure but clearly present. Defence projects can be similar too, in that specific benefits can be difficult to articulate accurately in monetary terms and yet can be worth everything in terms of future security.

However, regardless of the profile of the project being proposed, the chosen delivery route or option needs to be selected and justified. The possibility of both tangible and intangible benefits is articulated not just by the APM, but also by the PMI,[2] who define a business case as being a summary of a project detailing why a client should, or should not, undertake the project and one that can include both financial and nonfinancial benefits.

These definitions do not overtly recognise the 'through-life' nature of the business case. The Chartered Institute of Building[3] (CIOB), express the view that not only should the business case obtain and record commitment to the project, but should also provide a framework for the planning and the management of the project. This begins to introduce the purpose of the business case, both as a foundation block on which the project is built and delivered, and as a tool with a focus on the full duration of the project, thereby ensuring the project's ongoing validity and relevance to the sponsoring organisation.

DOI: 10.1201/9781003470748-3

The Japanese Project Standard (P2M[4]) incorporates the basis of the business case in a document called the Project Charter. It may be more accurate to say that certain items are added to the business case within the P2M Project Charter. The business case, once signed off though, serves a similar purpose in providing the official approval for the project to begin. The P2M approach to the project charter will also ensure that a project manager is appointed at this early stage and that organisational resources are available to the project.

The business case should therefore make certain that:

1 The goals of the project are clear and an overview of the time, cost and quality, or project benefit parameters, for the project are established.
2 The challenges and risks associated with the benefit delivery should be identified and considered along with any project constraints.
3 A framework for the planning and the management of the project is provided.
4 The ongoing validity of the project is continually assured on a through-life basis.

2.2 How Is the Business Case Delivered, Managed, and Used?

The first section of this chapter has shown that the main Project Management bodies agree that the business case should validate any proposed project and demonstrate that the project is worth doing.

In seeking to validate a project's selection and approval, there are two key questions that need to be answered: the What and the Why? The 'What' defines the project's end deliverables and the associated costs. The 'Why' describes the purpose, or benefits of delivering the project. In order to gain authorisation to proceed, a project must deliver more in terms of benefits than it costs to produce these benefits, as illustrated in Figure 2.1, the 'Why' must outweigh the 'What'. The project must deliver a net positive return to be considered worthwhile and to be seen as a credible option.

The APM[1] definitions of Project Governance point towards two key elements that need to be addressed within project completion, namely, efficient and effective project delivery. The business case reflects this. The costs must be delivered efficiently in a cost-effective manner, and the benefits must be effective so that they are of maximum use to the sponsoring organisation.

Simply leaving the question as it stands above though, is not sufficient. Greater granularity needs to be added if the right decision on the project's viability is to be made.

The What and the Why should be broken down and analysed in more detail. Without this work the project manager will have little idea of their project which would make it very difficult to gain the commitment of senior management and indeed other key stakeholders (see Chapter 6).

Figure 2.2 below shows that the What and the Why translate into Cost and Benefits. The types of costs and benefits involved in a project can vary widely between project types, but diagram below, at Figure 2.2, offers some possible examples.

Figure 2.1 The benefits of a project must outweigh the costs, the why? must outweigh the what?

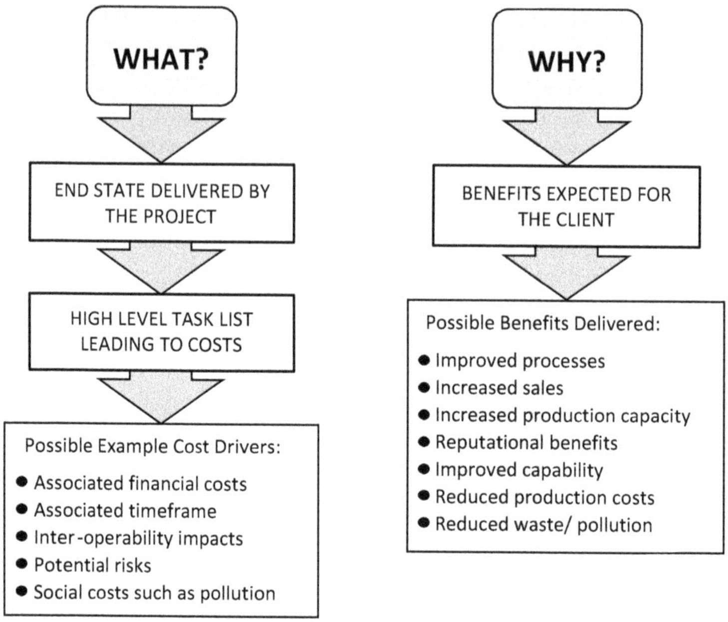

Figure 2.2 Possible generic costs and benefits within a business case.

Just as the range of costs can be many and varied, so can the nature of the benefits.

The diagram at Figure 2.2 represents a sample of potential costs and benefits but there will be many additional and alternative costs and benefits which could potentially be identified. One of the reasons for this is that projects can have significantly different purposes, not just in terms of what they deliver, but also in terms of who they deliver for. Chapter 4 on Project Lifecycles expands on this, noting that client expectations are considerably different depending on, for example, whether the client is commissioning the project with a view to use, or to sell, the project's output. In exploring the business case, what is important however, is that both the project costs and the financial worth of the benefits produced, can be identified and measured. An approach to managing this is reviewed in Chapters 10 and 12 respectively.

2.2.1 Other Than Costs and Benefits, What Else Should the Business Case Address?

While accurate estimates of the Cost and Benefits are important, so too is an assessment of how the project outputs will meet the purposes of the organisation for whom the project is delivered.

The business case therefore needs to address more than just the project costs and benefits. If consideration is to be given to the different elements that need to be included in the document, it is key to be clear about the purpose of the business case.

In considering the list at the end of Section 2.1, it is evident that the business case must:

1 Provide drive and direction for the project.
2 Be agreed as both robust and defensible before the project starts.
3 Remain robust and defensible throughout the project if the project is to continue.

The last of these three implies that the business case is a 'living document' which continues to be relevant and will hence need updating throughout the life of the project. It also hints at the use of the business case as a project baseline, in that if the business case is no longer relevant, if it no longer offers an excess of benefits over costs, the project should be closed. One important point in making this statement is that the costs referred to relate to the unspent costs of the project. The spent, or sunk costs are exactly that – sunk and cannot generally, with the exception of monies received for scrap, be recovered. The project should therefore continue if the benefits that will be delivered exceed the unspent portion of the costs to completion.

Once established and signed off, the business case demonstrates management approval and buy-in. It provides approval for funding and confirms the rationale for it. It constructs a stable platform on which the project can be delivered, and against which progress can be measured.

Having proved that the project is something that should be delivered, how though, can it be determined that the chosen delivery mechanism is the best option? In order to establish this, more information is needed within a business case than a simple cost benefit analysis, although of course this is a vitally important element that will help to justify decisions.

The business case needs to prove that the project is aligned with corporate strategy. It may be an excellent project, but is, for example, the construction of housing a suitable project for a road builder? Is the production and retail of new computer hardware a good project for a food manufacturer? The answer is almost certainly 'no' since each of the organisations undertaking the project have no experience in this new sector and will have been established to exploit the opportunities in their own, specialist sectors. To undertake either of these two projects would require the company in question to re-set their strategy.

Beyond the need to clearly identify the costs and benefits, the business case must also therefore establish the strategic need for the project and the benefits it will provide. It is important too that the desired end state for the project is clearly, and fully, described and to do this, the project team will need to ask a number of key questions such as: What does a successful outcome look like? How will the team know when the project is complete? How will this end state be reached? How will the various parts of the project be procured? How will these elements be funded? Once the benefits that the project will deliver have been determined, how will the project ensure that the organisation actually realises these benefits? Are the benefits certain, or are there risks associated with delivering the benefits? If there are risks, is it clear how these risks will be addressed?

Addressing these questions early in the project's lifecycle will add clarity to the project and help ensure that when the business case is communicated to the project team, the team is pulling in the same direction, helping it to deliver in an efficient manner.

2.2.2 *Business Case Contents*

The business case should, at a minimum, therefore, cover the following areas:

The first of the points in Figure 2.3 below focuses on the specifics of the project itself – what does the end state, after the project has been completed, look like, and what will need to be delivered in order to enable these goals to be met?

The Strategic Case will provide a detailed description of how the project will contribute to the objectives of the organisation, and how it will meet the business need. It will review the main stakeholders for the project and compare how the existing scenario and the post project delivery scenario will impact on them. The project scope will be addressed, and any constraints or dependencies considered, along with the project's critical success factors.

1. Project definition and objectives
2. The strategic case for the project – is it aligned to company strategy?
3. How will the project be delivered – options analysis.
4. The expected benefits (and dis-benefits)
5. The costs and funding arrangements – how affordable is the project?
6. The project's risk profile – how achievable is the project?
7. The project's timescale.

Figure 2.3 Outline contents of the Project Business Case.

The third item in Figure 2.3 – Option Analysis, or Option Appraisal, will be considered below in the Options Appraisal section where the methods used, and purpose of the appraisal will be discussed and analysed in detail.

Both the expected benefits and dis-benefits, or negative benefits, should be clearly laid out along with the risks to which any benefit delivery might be subject. This is normally done in monetary terms as most benefits have a financial value. Occasionally, as discussed above, it is sometimes difficult to put a monetary value on the benefits of a project. In such cases a Cost Effectiveness Analysis may be undertaken. In this instance, it is accepted that the project output is required, it may perhaps be a mandatory defence or regulatory requirement. Having accepted that the project is non-negotiable, the Cost Effectiveness Analysis centres on estimates of the net present costs required to provide the project output and puts to one side the value of the benefits which must be delivered. If the different methods produce variations in the outputs, then these will be as far as possible quantified using either monetary or judgemental methods. Suffice it to say however, the vast majority of projects that are come across will be such that their benefits can be expressed in monetary terms.

The fifth point from the list in Figure 2.3 considers the affordability of the project. This will include issues such as how elements of the project are sourced – in-house or out-sourced, and if the latter, from which supplier? This section can also include contractual conditions in terms of timelines, specifications or risk allocations and risk transfers. These issues lead to the consideration of the project's affordability. It will consider when cashflows are due into the projects bank account, when cash is due to leave the bank account and the amount of each of these flows. Just as they can for businesses in general, poorly managed cash flows can be a disaster for any project and can lead directly to overspends due to increased project borrowing requirements.

The mention of risks above leads onto discussion of likely areas of risk within the project. Similar projects, delivered previously, might be reviewed alongside various areas of the proposed projects. For instance, what are the risks involved in the chosen delivery strategy? What are the potential risks given the individuals in the team assembled to deliver the project? Do they have the time, skills and experience needed? What are the financial risks of the project? One area, mentioned above, that is often forgotten, is that of risks to the delivery of benefits. The reason the client is spending money on the project is that they expect to receive some benefits from its delivery. Just as important therefore as controlling the risks that could impact the costs is the ability to control the risks that could dilute or even destroy the project's benefits. This section of the Business plan should reflect this by including top level discussion of the risk management strategy and contingency plans as appropriate.

Larger more costly projects will require additional work as part of the business case development, yet the areas addressed by the content will be similar to those above. This approach helps to ensure that the depth of the appraisal provided is commensurate with the scale, complexity and even national importance of the project. Governments when constructing a Business Case for national projects will often include separate sections covering the strategic/ economic/ commercial/ financial aspects of the project. These sections will be similar in content to those above but are likely to be more detailed reflecting the increased scale of the project.

The Strategic Case will address why the change is required and will clearly define the outcomes and scope defined by the project's end state, and how the project will deliver the sponsor's strategic aims. The business case assesses the economic costs and benefits of the proposal to society as a whole, including the positive and negative externalities, and will cover the entire period of the project lifecycle. The Commercial Case seeks to demonstrate that the project will deliver a viable and well-structured procurement solution including both its planning and its management. The financial case needs to demonstrate that not just the initial investment or capital costs are affordable, but so too are any potential ongoing running costs of the project once delivered. These four aspects are generally addressed in all business cases, but in these very large projects may be addressed as a group of documents each addressing one of these elements.

Whether the business case is delivered as one document or multiple documents the function is the same, to provide reassurance, definition and confidence in the project's viability. The consideration and evaluation of these cases, will mean that choices have to be made relating to how the project is costed, procured and financed. Assuming there is more than one way of delivering the project, it will be necessary to select from a range of alternative strategies. The way in which this is done brings the discussion back to point 3 of Figure 2.3 above – 'Option Analysis'.

2.2.3 *Option Analysis*

Option Analysis, also known as Option Appraisal, is defined as a method of assessing project information in a structured way to enable informed decisions to be made between various delivery options. It allows the costs and benefits of different project solutions to be compared on a like for like basis, thereby demonstrating which offers the best value for money.

It is usually considered at the start of a project, as part of the business case, but it can help in assessing the desirability of alternative delivery methods, funding approaches, risk mitigation, or indeed any project option, throughout the lifecycle of the project. In doing so it not only helps the project manager to make a sound decision as to which option to pursue but also gives the assurance of a robust underpinning to the decisions made. The appraisal will show the costs and the benefits that the option is expected to deliver, thus providing both the baseline needed to enable the project to be monitored and controlled (see Chapter 11) and it will also list the expected benefits which can be used to inform the projects approach to benefit management (see Chapter 12). This can be useful too when it comes to the project review as a record of why decisions were made, particularly in terms of expected benefits.

There are three main methods of Investment Appraisal which enable the project manager to take estimates of the monetary value of both project costs and benefits and to compare them. They have a varying degree of sophistication, but each can provide the project manager with important information. These methods are Payback, Net Present Value, and the Internal Rate of Return.

2.2.4 *The Payback Appraisal Method*

The first and simplest of the methods is a calculation of the Payback period. A Payback calculation will calculate how long the project is operating in a negative cashflow position.

Suppose a small project (Project X) to introduce a cost saving to a process, is being considered. The project will need an initial outlay of £250,000 immediately (year 0), but by the end of the first year (year 1) will return £120,000. Table 2.1 below shows the remaining cashflows across the project's 4-year life. The calculations show the profit made across the years and the income from the disposal of the equipment in year 4:

Table 2.1 also shows the net cashflow for each year and the project's cumulative cashflow for over the 4 years is shown on the right-hand side. If the cumulative cashflow is plotted along a timeline, then Figure 2.4 shows the resulting graph:

Table 2.1 Example – Project X's Cashflows

Project X Cashflows		*£ (ks)*	*Cumulative Cashflow*
Year 0	Project's initial outlay	**−250**	**−250**
Year 1	Net income in yr 1	**+120**	**−130**
Year 2	Net income in yr 2	**+70**	**−60**
Year 3	Net income in yr 3	**+20**	**−40**
Year 4	Net income in yr 4	**+120**	**+80**

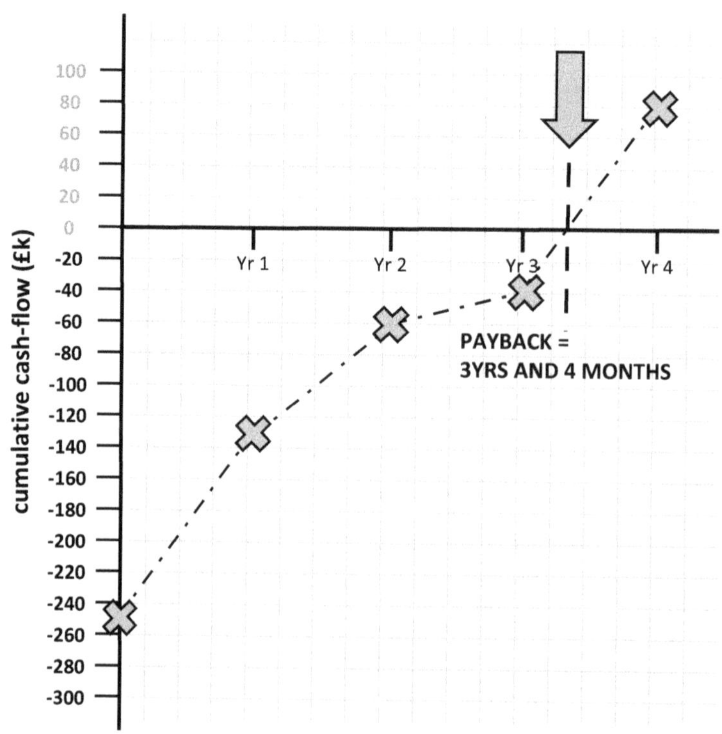

Figure 2.4 Project X's cumulative cashflows.

After 3 years the graph shows that the cumulative cashflow is −£40,000. It just needs an extra £40,000 or 40/120ths of year 4's cashflow to make the cashflow neutral, so the overall Payback period is 3 years 4 months, at which time the project sponsor has recouped the money which he has spent on the project.

As a measure it has a number of advantages. It is straight forward to understand and easy to calculate. The Payback method of Investment Appraisal can be used on its own, or alongside either of the other methods mentioned and it helps to determine how long the project is financially exposed before it starts to make money.

It is often said though that it has two main weaknesses. The first of these is that it only considers cashflow up to the breakeven point. Suppose two alternative projects, A and B, with identical deliverables, had the same cashflow as shown in Table 2.1, but they both extended into a fifth year. Suppose in year 5 Project A had a positive cashflow of £100,000 and project B had a positive cashflow of £1,000,000. Clearly all other things being equal, project B would be the better, but this would not be shown by the Payback period – both would have identical outcomes, breaking even after 3 years and 4 months! The weakness of only considering cashflows up until the breakeven point and not beyond, is clearly illustrated here.

Secondly, it is relevant to ask if the project would rather earn £100,000 today or £100,000 in a year's time? The answer is hopefully obvious – £100,000 now, but why? One factor is that the £100,000 could be invested, say at an interest rate of 10%, by the end of the year the £100,000 would be worth £110,000 which is clearly preferable to waiting a whole year to receive £100,000. Another reason why it could be better not to wait a year for the payment is related to risk. If a third party was going to donate £100,000 next year, there is a risk that in the meantime they might change their mind or worse still become bankrupt, hence the money given today is more certain and the offer of it is more valuable.

How can this variability in the time value of money be factored into investment appraisals and what impact would it have?

2.2.5 *The Net Present Value Appraisal Method*

The Net Present Value (NPV) approach to project Investment Appraisal takes estimates of the future net cashflows for a project delivery option and converts them back into today's monetary values. This then allows project options to be evaluated and decisions to be made on a like for like basis, no matter how different their cashflows or the timings of them are, which presents the project manager with the project equivalent of comparing apples with apples.

The first step in calculating an NPV appraisal is to determine the lifespan of the project and the number of years over which the project will generate cashflows of one kind or another. Once this is determined, then the project manager needs to calculate the net cashflow (any income less any costs), associated with each year. The discussion of the time value of money has already recognised that a sum paid in a year's time is of less value than if that same sum were paid today. How much less will depend on how many years into the future the sum is paid and what is termed the 'discount rate'.

The discount rate calculates how much the future value of money needs to be reduced by in order to equate to its value in terms of today's money. This discount rate can be chosen to reflect a variety of things, such as the current interest rates, the risk associated with these future payments or the cost of borrowing money. The discount rate should result in two figures-one in today's values and one in terms of a future date. The project should be ambivalent between the two, both the present value and future value should have equal value to the potential investor.

Having decided the discount rate, discount factor tables can be found in numerous places including on the internet. If a discount rate of 8% per annum was chosen, then the discount factor table for years 0–4 is shown in Table 2.2 below:

Table 2.2 Discount Factors at 8% for Years 0–4

Year	0	1	2	3	4
Discount rate at 8%	1.00	0.93	0.86	0.79	0.74

Table 2.3 Net Present Value at an 8% Discount

		£ (ks)	*Discount Factor at 8%*	*Discounted Cashflow £(ks)*
Year 0	Project's initial outlay	**−250**	**1.00**	**−250.00**
Year 1	Net income in yr 1	**+120**	**0.93**	**111.60**
Year 2	Net income in yr 2	**+70**	**0.86**	**60.20**
Year 3	Net income in yr 3	**+20**	**0.79**	**15.80**
Year 4	Net income in yr 4	**+120**	**0.74**	**88.80**
			NPV	**£26,400**

The table tells us that £1 at time zero (now) is worth 1.00 × £1, but if we were offered £1 in 3 years' time it would only be worth the same as 79p today (£1 × 0.79). An alternative way of looking at this is that if we invested 79p today at a rate of interest of 8% we would have £1 at the end of year 3. To calculate the value of any year 3 cashflow in today's terms we simply have to multiply it by 0.79, and similarly for the other years using the discount factor associated with each particular year.

Using the data for Project X shown in Table 2.1., Table 2.3 shows that by multiplying each year's cashflow by the relevant discount factor, the cashflow generated, or value, from that project can be calculated in terms of today's money – the 'discounted cashflow'.

The project therefore returns £26,400 in terms of today's money, which represents a return on the original investment of 26,400/250,000 = 10.5%.

Only values in the future are discounted. Money spent today is included at face value, hence the 'discount factor' of 1.00 in Year 0, shows that the value of outlay on day 1 of the project remains the same. The difference therefore between NPV and discounted cash is that NPV includes the initial outlay, the initial investment in the project. For that reason the tables shown here illustrate the NPV of the project rather than the discounted cashflow.

2.2.6 *Comparing Projects and Project Options Using NPV.*

The calculation in Table 2.3 shows that the project has the potential to be profitable. Is it the best option though to deliver the project outcomes defined in the business case? There could be a second option, for example, one that delivers the benefits over a 5 year period. The cashflow for this second, alternative, project is shown below in Table 2.4.

Following the same process shown in Table 2.3 above, Table 2.5 shows the discounted cashflows and the NPV for the 5-year option:

A comparison of Tables 2.3 and 2.5 shows, in terms of NPV, that the 5-year option generates a larger return, £41,590 than the 4-year option which generates a return of £26,400. These figures, despite the different lengths of the two project options, are directly comparable since all cashflows have been expressed in terms of today's monetary values. This example shows that NPV can be used to decide which of two or more project options is the best choice

Table 2.4 Project Cashflow – 5-Year Option

5 Yr Project Option		*£ (ks)*	*Cumulative Cashflow*
Year 0	Project's initial outlay	**−350**	**−350**
Year 1	Net income in yr 1	**+120**	**−230**
Year 2	Net income in yr 2	**+120**	**−110**
Year 3	Net income in yr 3	**+120**	**+10**
Year 4	Net income in yr 4	**+120**	**+130**
Year 5	Net income in yr 5	**−10**	**+120**

Table 2.5 Net Present Value at an 8% Discount Rate – Five-Year Option

5 Yr Project Option		*£ (ks)*	*Discount Factor at 8%*	*Discounted Cashflow £(ks)*
Year 0	Project's initial outlay	**−350**	**1.00**	**−350.00**
Year 1	Net income in yr 1	**+120**	**0.93**	**111.60**
Year 2	Net income in yr 2	**+120**	**0.86**	**103.20**
Year 3	Net income in yr 3	**+120**	**0.79**	**94.80**
Year 4	Net income in yr 4	**+120**	**0.74**	**88.80**
Year 5	Net income in yr 5	**−10**	**0.68**	**−6.81**
			NPV	**£41,590**

financially. This could be done at the outset of a project or part way through when a decision needs to be made as to, for example, whether a new technology which presents itself, should be used or not.

When choosing which investment option to take, the NPV is a key part of the decision, but not the only part. Aspects of the project such as its strategic fit, the level of risk associated, which might include a consideration of interest rate volatility, need to be taken into account. Organisations may consider how sensitive the project's cashflow is too, for example, how would a 2% fluctuation in interest rates either way impact the projects viability? The project example in Table 2.5 could be re-evaluated using not just an 8% discount rate but also a 6% and a 10% rate. This would reveal if the project was still profitable at these bracketed values and hence reveal how sensitive the project was to interest rate fluctuations.

Another way in which a company might take account of market volatilities could be to only invest in projects with a return above a certain level. This is the basis of an assessment technique known as the Internal Rate of Return, or IRR.

2.2.7 *The Internal Rate of Return (IRR) Appraisal Method*

Companies considering investing in projects will often have a view as to the level of return they need a project to deliver before they choose to invest in it. The examples above have shown that if a project has a positive return, at the chosen discount rate, it should be considered for investment. The flip side of this is that if the return were shown to be negative at a given discount rate, then, the project should not be invested in. In making this judgement companies will often have expectations as to how large the potential return needs to be before they choose to invest. This rate of return is often called a 'hurdle rate'. Any potential investment must offer a return higher than, or 'get over', the hurdle rate before it is considered. If it fails to make it over the hurdle rate, then the project will not be considered any further. The project's rate of return is calculated using a technique called the internal rate of return, or IRR.

The IRR is calculated in a way very similar to that used to calculate the NPV. Using exactly the same method as described above, the main difference being that the same project option is analysed using two different discount rates.

For example, taking the figures shown in Table 2.5 above, what would happen if a discount rate of 15% was used as well as the 8% rate used above?

Table 2.6 shows that at a discount rate of 8%, the NPV, the value of the project in today's terms, is positive (+£41,590). At a discount rate of 15% however, the NPV is negative (−£11,800). As the discount rate increases, the value of future positive cashflow becomes less when compared to the initial, Year 0, outlay, until a point is reached where the discounted cashflow turns from positive to negative. In the case of the project shown in Table 2.6, this would be at a point somewhere between a discount rate of 8% and 15%. At this precise point, the initial outlay would be exactly offset by the discounted cashflow and the NPV would be zero. This shows the interest rate, (represented by the discount rate) at which the project breaks even, or the point where the project makes neither a profit nor a loss.

This point can be mapped on a graph as shown in Figure 2.5 below and is known as the IRR:

Any company undertaking an investment will expect to get a positive return on that project. The size of return which the company expects may be set at a level of X% where X can vary according to the industry or the individual company's approach to investment. The rate of X%

Table 2.6 Net Present Value at a Discount Rate of Both 8% and 15%

5 Yr Project Option £ (ks) (@8%)			*Discount Factor at 8%*	*Discounted Cashflow £(ks)*	*5 Yr Project Option £ (ks) (@15%)*			*Discount Factor at 15%*	*Discounted Cashflow £(ks)*
Year 0	Project's initial outlay	**−350**	**1.00**	**−350.00**	Year 0	Project's initial outlay	**−350**	**1.00**	**−350.00**
Year 1	Net income in yr 1:	**+120**	**0.93**	**111.60**	Year 1	Net income in yr 1:	**+120**	**0.87**	**104.40**
Year 2	Net income in yr 2:	**+120**	**0.86**	**103.20**	Year 2	Net income in yr 2:	**+120**	**0.76**	**91.20**
Year 3	Net income in yr 3:	**+120**	**0.79**	**94.80**	Year 3	Net income in yr 3:	**+120**	**0.66**	**79.20**
Year 4	Net income in yr 4:	**+120**	**0.74**	**88.80**	Year 4	Net income in yr 4:	**+120**	**0.57**	**68.40**
Year 5	Net income in yr 5:	**−10**	**0.68**	**−6.81**	Year 5	Net income in yr 5:	**−10**	**0.50**	**−5.00**
			NPV	**£41,590**				**NPV**	**−£11,800**

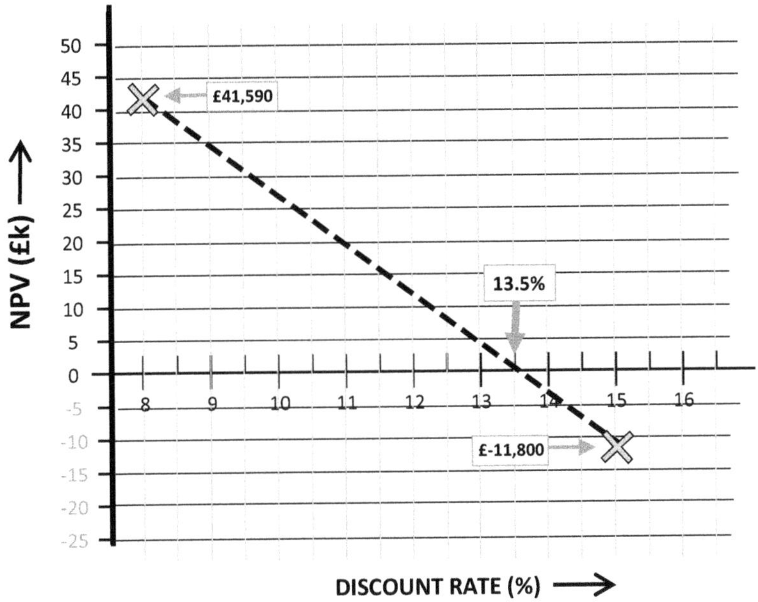

Figure 2.5 Graphical calculation of the IRR.

is, as mentioned, known as a hurdle rate. In Figure 2.5, if the hurdle rate is less than 13.5%, for example 10%, then reading off the graph, the project offers a return of approximately £27,000 when future cashflows are discounted by 10%. If, however, the hurdle rate is greater than, or to the right of 13.5%, then the project will return a loss as illustrated by the graph which shows a negative figure of approximately −£4,000 at a discount rate of rate of 14%. The company would on the basis of the IRR (13.5% in Figure 2.5), choose not to invest when it's hurdle rate is greater than the IRR, the implication being that funds are limited, and higher returns can be made elsewhere. The hurdle rate can be set to reflect multiple factors, it is quite often calculated as:

Cost of Capital + Risk Premium = Hurdle Rate

Projects that have a higher risk profile can therefore be given a higher 'hurdle' to overcome before they are accepted.

2.2.8 *Summary of Investment Appraisal Methods*

Each of the three methods discussed have some similar weaknesses. All three, Payback, NPV and IRR rely on estimates of future income which can vary in accuracy (see Chapter 7). Payback however, also ignores the time value of money and the cashflows past the Payback point, while the specific discount rate used within an NPV calculation could be subjective, as indeed could the setting of the hurdle rate within an IRR calculation. The use however of a common rate between calculations does enable comparisons to be made and although it can be subjective, the rate can be chosen to reflect the organisations view of prevailing market conditions.

The three methods can be used together, and although the Payback method might be considered the weakest of the three, it's strength can be seen when used in partnership with one of the other two methods. In this scenario, Payback can reflect the risk inherent in a project. The return could be calculated by using the NPV approach, while use of the Payback method would reveal how long the project had a negative cumulative cashflow balance and therefore how long it was financially exposed – the longer this is then the greater the level of risk is within the project.

2.2.9 When Could Investment Appraisal Be Used and for What Purposes?

Whichever method of Investment Appraisal is used, there are multiple points and purposes during the project lifecycle (see Chapter 4,) where its application can add benefit and help the project manager make the right decision.

It can be used at the end of each project stage to confirm the project's ongoing viability and determine its readiness to progress. It can also be used at any point throughout the project lifecycle where there are decisions to be made about which project option or delivery method to choose. This could be related to for instance, the risk mitigation process, the assessment of change control requests or the assessment of alternative build options. These may present themselves at any point during project delivery, so Investment Appraisal is a useful process throughout the project lifecycle. Of course, given the importance of cost and budgets to the control of a project, it is an essential tool for the project manager during the drafting of the business case, but it can provide far more than simple financial analysis for the project.

It goes without saying that the business case provides a Cost v Benefit Analysis as shown in Figure 2.2 above. A benefit could be described as a financial outcome, or as an output, or an ability that the project provides, multiplied by the amount of use that can be obtained from it. Simplistically a benefit could be described as a quality delivered multiplied by the amount of time that it can be used for.

What is important though is that the business case defines the outputs, qualities or benefits that the project is targeted with delivering. It should define the end state and how it is reached. This is vitally important in terms of project communication. It is key in enabling the project team to have a common understanding of the project so that the team can all pull in the same direction. Once agreed, this certainty around the project goals adds considerably to the efficient and effective delivery of the project. The business case also serves as an initial baseline against which the project's progress can be measured, and the impact of potential changes measured. The impact on a project of not establishing a sound business case can be seen in the next part of this chapter where the delivery of the Millennium Dome project is considered.

2.3 Project Review: The impact of the Business Case on the Project: The Millennium Dome

2.3.1 The Project Background

The Millennium Dome now renamed and repurposed as the O2, was built in Greenwich in London at the end of the 20th century to house an exhibition to celebrate the approaching 21st Century and the 3rd Millennium. Having been first discussed in 1996, the construction itself began in 1998 with the foundations being completed along with the main supports and the Teflon coated roof. It's sponsor or client was the New Millennium Experience Company Ltd (NMEC) formed by the then Labour Government led by Tony Blair. The Encyclopaedia Britannica[5] states that the Dome is the largest 'Dome' in the world, having 80,000 square metres (over 861,000 square feet) of floor space inside, and is almost twice as big as the second largest Dome, the Georgia Dome in Atlanta in the US. On a technical note, the description of it as a Dome, is contested by the Institution of Civil Engineers (ICE),[6] who say that in its true sense, the Dome is not a Dome because it is not self-supporting but is held up by 2,600 cables suspended from 12 steel towers 100 metre high, despite this, the building was universally known as the Millennium Dome when built and is now known as the O2.

2.3.2 *The Project Issue*

The purpose of the Dome as described by the National Audit Office,[7] was to provide a 'Millennium Exhibition of a scale and stature comparable with that of the Great Exhibition of 1851 and the Festival of Britain of 1951'. The project however, experienced 'severe financial difficulties' and required an additional £175m of lottery funding to keep it on track. The cashflows from receipts too, were less than projected, as a result of visitor numbers being 'substantially lower than the 12 million envisaged in the plan'.[7] This was the subject of questions in Parliament[8] where the 'reality of a plan which saw the Dome attracting four times as many visitors as the next most popular UK pay-to-visit attraction' was questioned and shed some light on the veracity or strength of the business case produced. It was pointed out, in parliamentary debates, that it was considered by the private sector, before the start of the project that the 'risk outranked the reward'.[9] This perhaps should have called into doubt any approval of a business case at this stage.

In May 1997 the Labour Government came to power and, in line with conditions linked to their earlier support of the project, they undertook a review of the project in June 1997. As part of their review, the NMEC submitted to the Government and the Millennium Commission who had responsibility for providing, amongst other things, the National Lottery funding for the project, what was termed a revised business plan (not a revised business case.) The three main objectives of the plan were to

- create a world profile for the celebration of the Millennium in the UK
- deliver the exhibition and national programme to time and budget
- involve, engage and transform the visitor and participant.

The overall cost identified was shown as £758m of which £530m were build costs (excluding site acquisition and remediation costs) and £228m was linked directly to running costs. The financial benefits of the project, in terms of income, were stated as £359m. It was anticipated that at the end of the exhibition the Dome would be taken down and the site sold for £15m. It is unclear if the demolition/removal costs were included in these figures. The shortfall of £384m (£758m − (£359m + £15m)) was to be made up from grants. It was assumed in the plan that within the £359m of income, sponsorship would contribute £175m of direct income while the income from the estimated 12 million visitors through ticket sales, catering and merchandising would generate £169m.

A 1997 review of the plan did express concern that areas of the plan were lacking in detail, specifically on commercial, operational and pricing strategies and in addition there was little meaningful information relating to the content of the Dome, which would of course be key in attracting the expected 12 million paying visitors. The focus of the incoming government was on the future of Britain. This led to the emphasis of the Dome's contents being focused wholly on the future, thus neglecting to build on the foundations of displays relating to Britain's past 2000 years. This approach resulted in the criticism that the project was based not on efficient planning but too much on political ideas and figures.[10] Indeed, it is noted in the NAO's report that the business plan 'was only a rudimentary basis against which to assess content attractiveness and cost and whether the visitor projection of 12 million was correct' (Approximately only 6 million, of which 4.5 million were paying,[17] visits were eventually made to the Dome before the exhibition closed.)

The lack of direction and definition of purpose with regard to the project's contents is evident in a number of documents. It has been said that the "why" or vision of the project wasn't properly defined and that the project lacked planning and some elements such as the scope were unclear when the construction began. One source[11] stated that the Dome could be criticised for a 'total lack of an overarching vision', something which a properly constituted business case would have given it.

2.3.3 *How Could a Developed Business Case Have Helped?*

If the section above illustrates how the writing of the business case was approached, what then were the potential short comings and what was the impact on the project?

If the information presented above is contrasted with the content of the business case outlined in Figure 2.3 above, and repeated at Figure 2.6 below, it can be seen that there are areas for concern.

Of the seven key areas noted in Figure 2.6, it is probable that six of them are areas where the business case should add clarity, but could, in the case of the Dome, be questioned.

The one exception, and the one area of which there could be little doubt, was the project's opening date or timescale. Although the opening date was clear, even the overall project timescale could be questioned if the disposal element of the project lifecycle was considered. It was unclear at the start what would happen to the Dome after the exhibition. Disposal plans and costs were unclear, and at one time it was even muted that the Dome be relocated to Swindon[12] after the exhibition closed! The newspaper article reporting this referred to the '£800m structure on the Greenwich peninsula' as 'a costly white elephant', making it clear that there was a degree of angst over the future of the Dome declaring that 'ministers agonised over what to do next with what critics called a monument to government ineptitude.'

The project certainly was reported as lacking in definition, strategic alignment, clear delivery strategy, accurate benefit calculations and risk assessment. The costs were fairly well controlled, but the lack of income meant that cashflow was adversely affected.

It is reasonable to say that the lack of focus and definition within the business case was a key cause of the 'ineptitude' reported.

1. Project definition and objectives
2. The strategic case for the project – is it aligned to company strategy?
3. How will the project be delivered – options analysis.
4. The expected benefits (and dis-benefits)
5. The costs and funding arrangements – how affordable is the project?
6. The project's risk profile – how achievable is the project?
7. The project's timescale.

Figure 2.6 Outline contents of the Project Business Case.

The NAO report[7] expands on this 'ineptitude' referring to 'severe shortfalls in revenue', lower than predicted numbers in terms of both sponsorship revenue and visitor numbers. The estimates of visitor numbers were considered to be riskily high by a number of bodies, but remained in the forecasts, meaning that the expected benefits were overstated. The lack of receipts meant that the NAO report recognised that it was necessary, given the financial difficulties, to rely on receipts from the planned sale of the Dome. It was planned to sell the Dome to a company called Dome Europe plc, but they withdrew from any potential deal in September of 2000 leaving even greater uncertainty around the future sale proceeds. Unquestionably, the plans around the costs and funding arrangements could have been further developed within the business case, as could the degree of certainty about the Dome's contents, which a BBC report[13] said constrained the effectiveness of the Dome's marketing and hence its income. The lack of definition surrounding the contents or 'zones' of the Dome had drawn criticism for some time.

A House of Commons research paper as far back as 1998 reviewed the projects aims and stated 'However inspirational the Dome, the impact of the Experience will depend principally upon what is inside the Dome in 2000. It was on this most important of topics that we found official witnesses to this inquiry least informative.'[14] A newspaper report from 2019,[15] looking back at the Dome's performance recalls that 'the contents were 'panned'. They were described as underwhelming, compromised and communicating nothing in particular.' This lack of planning and knowledge of the contents points to a poor *Project Definition* element of the business case.

The *Strategic Case* for the project was equally difficult to define, although it was perhaps more understandable, given the focus on the provision 'of a centrepiece for the nation's Millennium celebrations' – a once in a thousand-year event, (although if viewed in the round, there were parallels with the 1851 Great Exhibition and the 1951 Festival of Britain). The Millennium Commission itself was established in February of 1994, and its policy was to fund Millennium celebrations including a possible national festival, thus perhaps hinting at a strategy. The possible 'national festival' turned out, following a 1995 consultation paper, to be the Millennium Dome.

An *Option Analysis* was performed with alternative sites (the shortlist of locations consisted of Birmingham, Derby, Stratford and the Greenwich site,[15] with the final two being the NEC in Birmingham and the site at Greenwich.) considered and alternative delivery mechanisms evaluated.

Originally private sector investment was sought to run the venture, but this proved impossible to facilitate. The NAO report[7] states the private sector would not accept the risks. Some of these risks would have been generated by the uncertainty and lack of definition within the business case, particularly that surrounding the nature of the content and hence the visitor numbers and therefore the resulting profitability of the venture. Instead, the Dome was financed by a company owned by the public sector, The Millennium Experience Company. The 12 zones within the exhibition were financed in part by commercial enterprises such as Ford and BT, which inevitably brought in commercial concerns that were not necessarily aligned with the overall strategy of the Dome such as it was. The sponsorship therefore helped address a financial shortfall, but arguably at the expense of the original concept. It has been stated that the 'Corporate sponsors generally exerted enormous influence over the design of contents in one way or another'.[16] This led to confusion and disagreements over what was wanted in terms of the content meaning that content designs were begun and then terminated and started again. This reduced the time available to complete the exhibition content and caused costs to rise leading to a large proportion of the project's contingency budget being spent on the content of the exhibition.

2.3.4 Conclusions on the Millennium Dome Project

The financial difficulties experienced by the Dome are widely recognised. The head of the NAO noted that the targets for visitors and income were 'highly ambitious and inherently risky.' This resulted in the project's grant funding needing to increase by 57% from £399m to £628m. The project itself has been the subject of a number of academic papers. The view presented in multiple studies is reflected particularly well in one of them.[17] It is said that from the very start of the project, its success was undermined. The nature of the project, a political endeavour was partly responsible for this due to the change of sponsor when the Labour Government came to power in 1997 and the project vision was realigned. The failure of the project was judged by the NAO as having its roots in the structure and focus of the project,[7,17] both aspects which should have been clarified and agreed within the business case. In particular, the project did not have clarity as to how the exhibition would be funded and delivered nor what role or influence the sponsors would have in this. The varied and uncoordinated approach to and from the sponsors led to the exhibition having a confused message which made it more difficult to promote. Additionally, the project's overall vision can be seen as weak and inconsistent which caused coherent communication to those involved with the project to be difficult at best. The lack of documented requirements and success criteria inevitably meant that different team members had different views on these issues. This in turn would have introduced friction to the project preventing efficient and effective project delivery. The lack of documented outcomes that should be part of any business case, ought to have begun to ring alarm bells and caused the project manager, to raise questions such as 'does the project have top management support?' 'Is the strategic need clear and obtainable?' and 'Will it get the resources it needs given the lack of defined outcomes?' Interestingly the continued delivery of the Millennium Dome project was questioned on a number of occasions, but it endured, due at least in part, to the reputational risk to the UK of it failing.[7]

2.4 The Business Case Before the Birth of Modern Project Management*

(* In this section, the research relating to the construction of HMS Barham 1912–1915 mentioned in the introduction is drawn upon to help illustrate how the subject of this chapter has developed.)

Looking further back into history prior to the Dome's construction, the start of what is termed Modern Project Management is widely attributed to either the Manhattan Project of the mid 1940s or the US Polaris Project in the late 1950s. Either way, there were clearly a vast number of projects delivered before these! One of the most interesting areas to search for these projects and what they can say about, in this instance what would now be termed the business case, is the construction of the Dreadnought ships of the First World War. They were large, expensive projects that were strategically important and funded by the public purse and hence detailed records were kept. These records, or what survives of them, are now held in various locations such as the National Archives, The National Maritime Museum and, having been built on the Clyde, the Scottish Business Archives held at Glasgow University.

Analysis of the records shows that the ships were built in a group of four or five vessels, all of one type or 'class'. The initial or 'first of class' ship was costed, and the total programme costs were discussed and approved in Parliament as part of the debate on the 'Naval Estimates' for each annual shipbuilding programme. Project definition would have been quite straight forward, around the turn of the century, the Two Power Standard determined the acceptable size of the British Fleet. This standard was defined by Earl Cawdor in the House of Lords, quoting the Prime Minister of the day, as being 'a preponderance of 10 per cent over the combined strengths

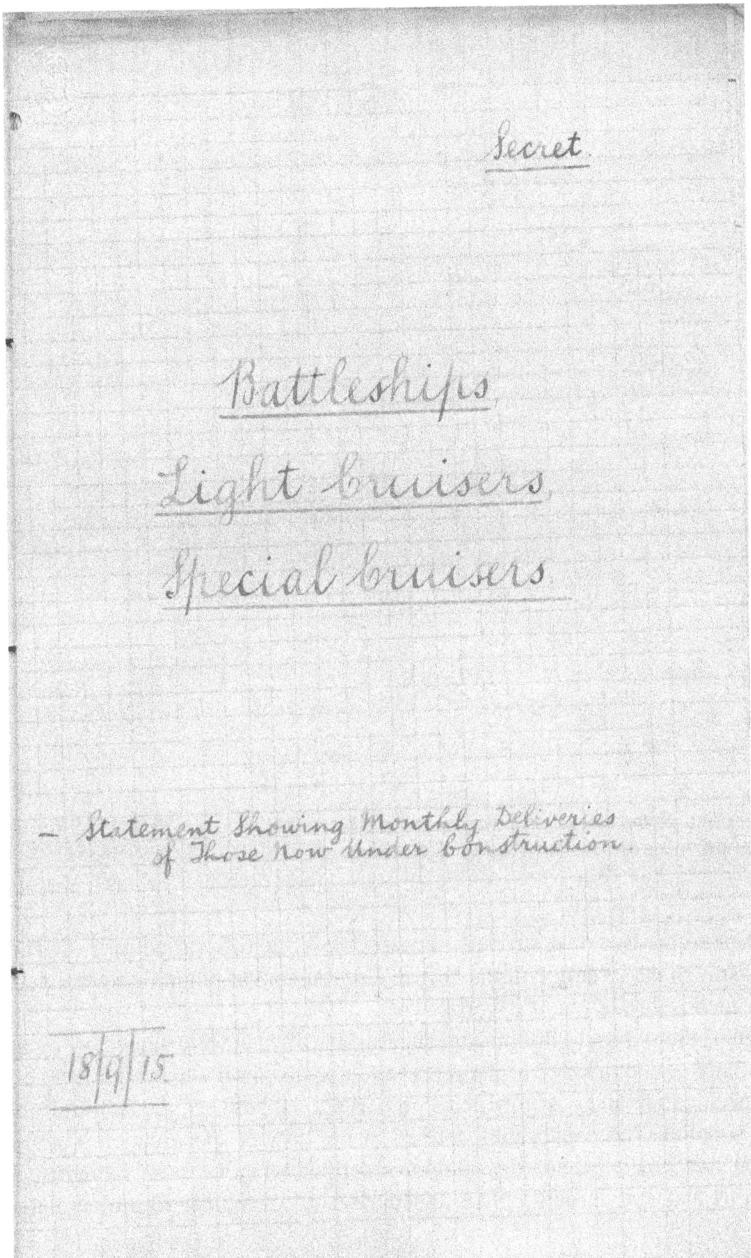

Secret.

Battleships.
Light Cruisers.
Special Cruisers.

– Statement Showing Monthly Deliveries of Those now Under Construction.

18/9/15

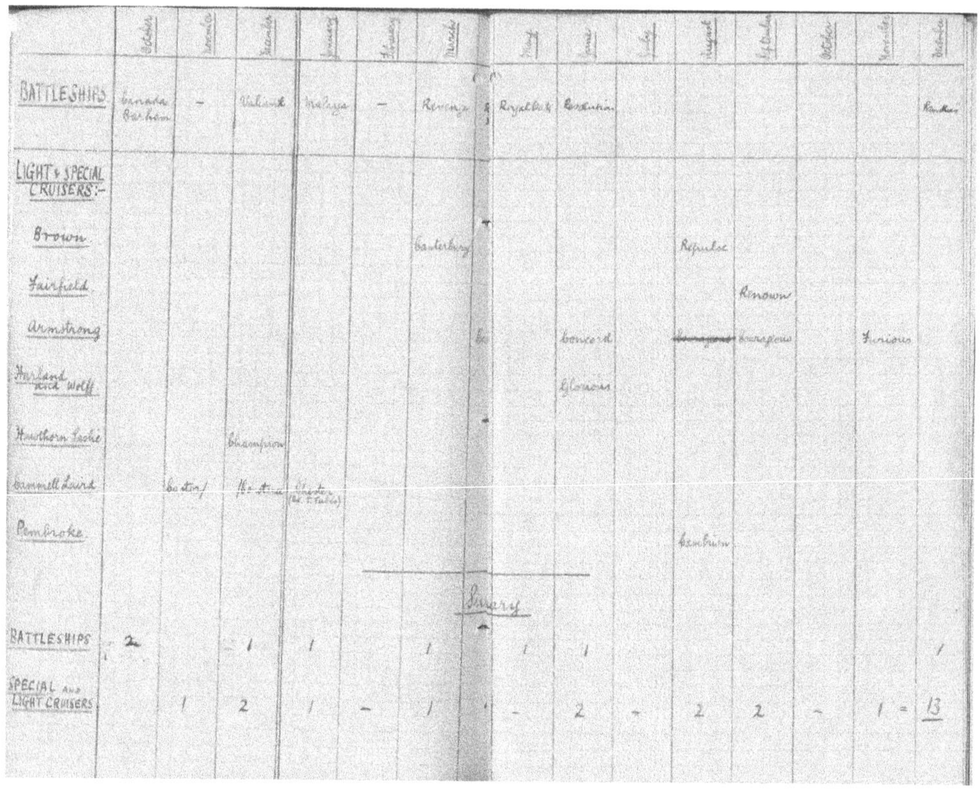

Photo 2.1 Report cover sheet and projected delivery dates as at 18/9/1915 for HMS Barham. By permission of The National Archives of the UK: ADM 1/8435/297.

in capital ships of the two next strongest Powers'.[18] While this varied over the coming years it was still very much the intention of the British Government to maintain fleet superiority in relation to other navies throughout the world.

The design of the ships and the ensuant capabilities of each were signed off by the Admiralty, as the client, and communicated to the relevant dockyard. Following high level cost estimates, a delivery specification and cost for each ship was agreed between the Admiralty and the dockyard as the supplier. The capabilities defined in the project specification, combined with the impact of the Two Power Standard would have defined what we today recognise as the project *benefits,* in this instance, largely linked to security and the ability to protect national interests of the time.

In respect of the Queen Elizabeth Class of Dreadnaughts, there is evidence of at least two different initial designs being considered, and each option being appraised as would be expected within a business case of today. With the imminent threat of war though, the 'do nothing option' was not included in the assessment! There is evidence of a *timescale* for the delivery for each ship. The dates shown in Admiralty documents would have been forecast in collaboration with the shipyard manufacturers and these are apparent from the initial contracts and in reports of the time throughout the project lifecycle. These dates, for each individual ship in Elizabeth Class, are shown in the report above (Photo 2.1). The report being dated 18 September 1915, shows the individual delivery dates as projections or forecasts between 1915 and December 1916. This can only be thought of as vital information in terms of predicting the rate of possible replacement against the rate of attrition of vessels during WW1.

Photo 2.2 shows an early document dated 22/10/1912 which has some examples of option analysis as well as a forecast delivery date for HMS Barham, one of the Elizabeth Class Dreadnoughts. The options considered in this document show that boilers from both Yarrow and Babcock were contemplated for the ships while the delivery date for the ship is given, in the bottom

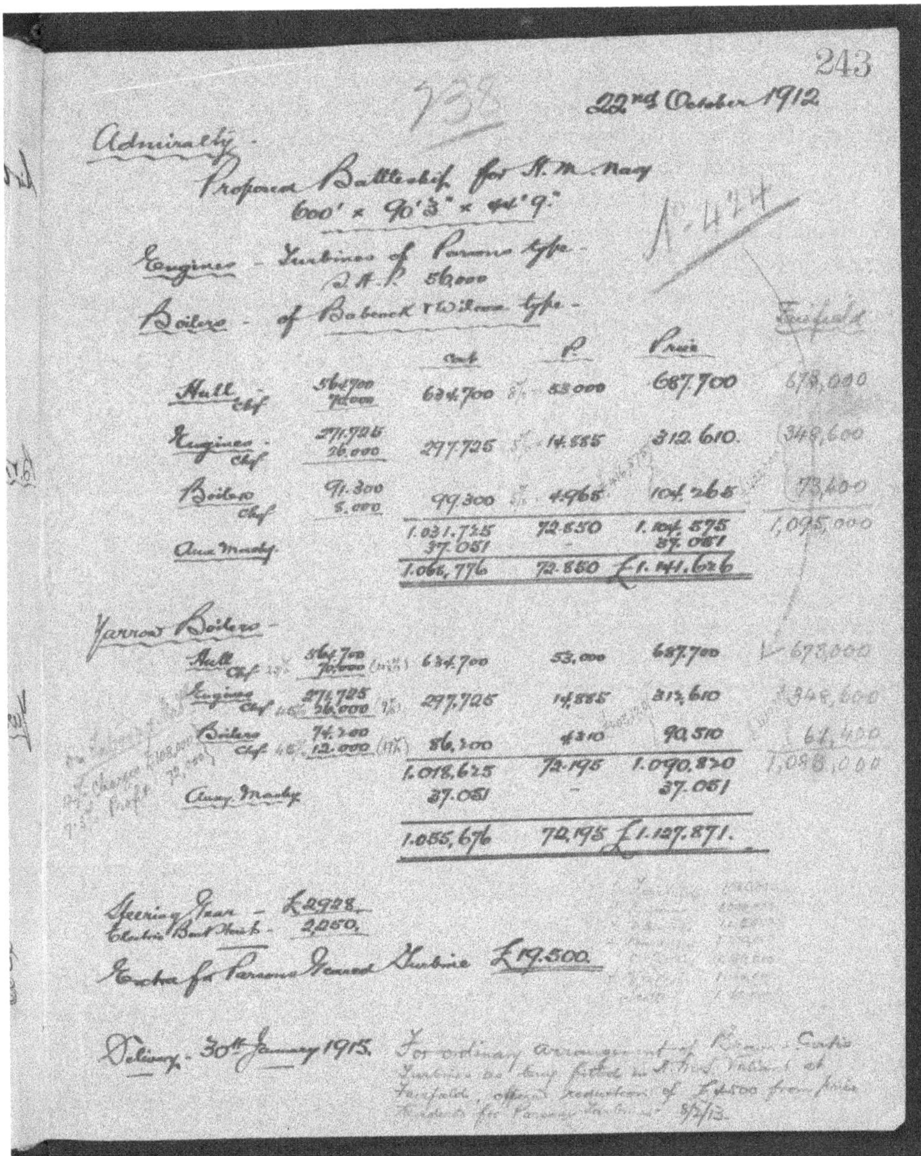
243

22nd October 1912

Admiralty.

Proposed Battleship for H.M. Navy

600' × 90'3" × 44'9"

Engines – Turbines of Parsons type

S.H.P. 56,000

Boilers – of Babcock & Wilcox type

		Cost	P	Price	Fairfield
Hull	564.700 / 70.000	634.700	53.000	687.700	678,000
Engines	271.725 / 26.000	297.725	14.885	312.610	348,600
Boilers	94.300 / 5.000	99.300	4.965	104.265	73,400
		1.031.725	72.850	1.104.575	1,095,000
Arm. Mounting		37.051		37.051	
		1.068.776	72.850	£1.141.626	

Yarrow Boilers –

		Cost	P	Price	Fairfield
Hull	564.700 / 70.000	634.700	53.000	687.700	678,000
Engines	271.725 / 26.000	297.725	14.885	312.610	348,600
Boilers	74.300 / 12.000	86.300	4.310	90.510	61,400
		1.018.725	72.195	1.090.820	1,088,000
Arm. Mounting		37.051		37.051	
		1.055.676	72.195	£1.127.871	

Steering Gear – £2928

Electric Boat Hoist – 2050

Extra for Parsons Geared Turbine £19.500

Delivery – 30th January 1915.

Photo 2.2 Early signs of option analysis (22/11/12) and delivery date (30/1/15) for HMS Barham. By permission of the National Records of Scotland and University of Glasgow Archives & Special Collections, Records of Upper Clyde Shipbuilders Ltd, Clydebank Division collection, GB248 UCS 1/74/6.

left-hand corner, as the 31st January 1915. The construction of these timelines would have had to be based on individual records and reports which have not survived the passage of time so well.

All elements listed in Figure 2.6 are therefore included in the review above, with the possible exception of the project's risk profile. As a development of the 1911–1912 shipbuilding programme the risks of construction would have been largely known. Additionally, with the 1st in Class being built in HM Dockyards and then subsequent vessels being constructed by commercial builders such as John Brown, Fairfields and Armstrong Whitworths, the risks inherent in the build designs themselves were encountered first in HM Dockyards before being transferred to commercial yards meaning that designs could be adapted if necessary, before the commercial projects commenced. Of course, each project was measured against the over-riding, but unspoken, risk of not being ready for the impending conflict, and hence the political risk of non-delivery overshadowed all other potential risks.

Today's project management approach offers advantages over that described above. It presents all the facets of the business case together in one document, adding clarity to the project and acting as an important communication and onboarding tool for any new project team member. However, the Dreadnought case underlines that the business case and its contents remain the foundation for any successful project.

References

1 APM. (2012). *APM Body of Knowledge* (6th ed.). Association for Project Management.
2 Project Management Institute. (2021) *The standard for project management and a guide to the project management body of knowledge* (7th ed.). Project Management Institute, Inc.
3 The Chartered Institute of Building (CIOB). (2014). *Code of practice for project management for construction and development*. John Wiley & Sons.
4 Ohara, S. (2017). *P2M – A guidebook of project & program management for enterprise innovation* (3rd ed.). Project Management Association of Japan (PMAJ).
5 Britannica. (2024). The Editors of Encyclopaedia. "Millennium Dome". *Encyclopedia Britannica.* https://www.britannica.com/topic/Millennium-Dome [Accessed 5 June 2024].
6 Institution of Civil Engineers. (2024). *Millennium Dome Greenwich, United Kingdom.* https://www.ice.org.uk/what-is-civil-engineering/infrastructure-projects/millennium-dome [Accessed 6 June 2024].
7 The Comptroller and Auditor General. (November 2010). *The Millennium Dome* [online]. The House of Commons. https://webarchive.nationalarchives.gov.uk/ukgwa/20170207052351/https://www.nao.org.uk/wp-content/uploads/2000/11/9900936es.pdf [Accessed 6 June 2024].
8 HL Deb. (29 November 2000). vol 619 cc1367–458. https://api.parliament.uk/historic-hansard/lords/2000/nov/29/millennium-dome [Accessed 19 August 2010].
9 Hansard HL Deb. (29 November 2000). Column 1379. https://publications.parliament.uk/pa/ld199900/ldhansrd/vo001129/text/01129-15.htm [Accessed 6 June 2024].
10 Akinradewo, O., Akinshipe, O. and Aigbavboa, C. (2022). Construction project failures around the world: What have we learnt so far? In 3th International Conference on Applied Human Factors and Ergonomics (AHFE 2022). https://doi.org/10.54941/ahfe1002301
11 Gray, C. (2003). The Millennium Dome: 'Falling from grace'. *Parliamentary Affairs*, *56*(3), 441–455.
12 Grierson, J. (30 December 2022). Newly released files reveal plan to move Millennium Dome to Swindon. https://www.theguardian.com/uk-news/2022/dec/30/newly-released-files-reveal-plan-move-millennium-dome-o2-swindon [Accessed 7 June 2024]
13 The BBC. (1 February 2002). Government's Dome handling attacked. https://news.bbc.co.uk/1/hi/uk_politics/1795133.stm [Accessed 5 June 2024]

14 The House of Commons. (12 March 1998). The Millennium Dome. Research Paper 98/32. https://researchbriefings.files.parliament.uk/documents/RP98-32/RP98-32.pdf [Accessed 5 June 2024]
15 Moore, R. (1 December 2019). The Millennium Dome 20 years on… revisiting a very British fiasco. https://www.theguardian.com/artanddesign/2019/dec/01/millennium-dome-20-years-on-new-labour [Accessed 3 June 2024]
16 Mcguigan, J. and Gilmore, A. (2002). The millennium dome: Sponsoring, meaning and visiting. *International Journal of Cultural Policy*, *8*(1), 1–20. https://doi.org/10.1080/10286630290032486
17 Alderman, N., and Ivory, C. (2011). Translation and Convergence in Projects: An Organizational Perspective on Project Success. *Project Management Journal*, *42*(5), 17–30. https://doi.org/10.1002/pmj.20261
18 *Hansard* HL Deb vol 197 cols 25–65 (November 1908).

Chapter 3

Project Feasibility and Sustainability

It may not necessarily seem obvious to cover project feasibility and project sustainability in the same chapter, yet the definitions given by the professional bodies for the two areas reveal strong connections between the two. Both concern themselves with the deliverability of benefits from the project, but sustainability takes a longer view in respect of when these benefits are delivered and exactly what should be included within the net benefits of a project.

3.1 The Purpose of Feasibility and Sustainability Within Projects

The PMI[1] describe feasibility as determining 'if the business case is valid and if the organisation has the capability to deliver the intended (project) outcome (benefits).' Feasibility is described as the first phase within the project lifecycle.

Feasibility therefore tends to be an inward-looking review, asking if the business case (see Chapter 2) delivers net benefits to the organisation and whether the organisation has the ability to deliver these benefits. Sustainability on the other hand, provides a more outward looking view of the project's impact. The APM[2] describe sustainability as an approach that encompasses, and seeks to balance, the environmental, social and economic impacts of a project. It does not ignore the needs of the current project stakeholders but focuses on a way of meeting their needs without compromising or overburdening future generations of stakeholders. The PMI[1] highlight that sustainability is a key dimension of the quality demonstrated by a project's output. As the APM's definition highlights, the PMI also asks whether the project deliverables impacts positively on economic, social and environmental parameters. This involves consideration of how the project is planned, particularly in relation to its use of materials and natural resources. P2M[3] makes the point that feasibility goes beyond the consideration of whether or not the correct goals are chosen and alongside this asks whether or not these goals are actually achievable and ultimately whether or not the project plan can succeed. P2M[3] also reports that companies are expected to align themselves with Corporate Social Responsibility (CSR) principles, which require an organisation to focus not on profit alone, but to focus also on helping to achieve a sustainable society by assuming responsibility for social and environmental issues.

The two concepts of profit and sustainability are not without conflict, the nature of which depends on the definition of profit used. Usually, a short-term definition is used, such as that utilised by accounting records which are reviewed at shareholder meetings which may help determine the immediate future of a company or its executives. If a longer-term definition of 'profit' is used, then this leaves room for elements such as accountability and social responsibility to be discussed within the concept of sustainability.

DOI: 10.1201/9781003470748-4

3.2 How Is Feasibilty and Sustainability Delivered Within Projects?

Having undertaken a project brainstorm to find possible project delivery options, a project team will be faced with a number of different possibilities, which will need to be narrowed down. One important question that must be addressed is how realistic or feasible each of the proposed solutions are.

3.2.1 *Project Feasibility Studies*

Projects are often described in terms of their time, cost and quality parameters. These three aspects were first highlighted by Barnes in 1969 who referred to them as the 'Iron Triangle' of PM constraints.[4] The triangle describes how the triple constraints of time, cost and quality interact and impact each other, being traded off against each other as the project develops in order to keep the project on track and maintain a feasible outcome. Thus, consideration of the achievability of these three project limitations can readily be seen as a measure of the project's feasibility. Other aspects though could also be considered. Quality itself is a measure of the worth of the project's output and is delivered once the project deliverables are ready to be used. As a result of this, it reflects a project lifecycle rather than an extended project lifecycle approach (see Chapter 4). If an extended lifecycle view of the project is taken, then the project manager is more accurately confronted with trade-offs between not time, cost and quality, but time, cost and benefits, as the delivered qualities are utilised and produce business benefits. It is interesting to note that the concepts of project lifecycles have developed since 1969, one of the changes being the introduction of the extended project lifecycle as an approach.

Rather than reflecting purely on time, cost and quality/ benefits, consideration of additional areas such as economic, cultural, safety and operational, technical feasibility and the legal implications of project delivery, will help give a more rounded picture of project feasibility. Note however, that each of these additional areas directly or indirectly support the delivery of one of more of the time, cost, quality/ benefits constraints of the Iron Triangle.

The precise areas that should be considered when addressing project feasibility can differ in specific projects, but the list given in the paragraph above is a good, relatively generic, starting point. If a project is subject to any particular pressures, such as schedule or delivery timescales, then these headings too should be included in the feasibility study. Failure to consider any one aspect of a project's feasibility could result in project overruns or even project failure, evidenced by cost and time overspends or potentially the early termination of a project. Each of the aspects of project feasibility discussed in the paragraph above are expanded upon below and will help to ask questions of the project.

The economic feasibility of the project is a key consideration. It is not just a question of whether the project will make a profit, or whether there is an excess of benefits over costs as described in the business case (see Chapter 2), but it is also a question of whether or not finance will be available to fund the project throughout its stages. If the project is relying on new technology, it should question whether the cost implications of the technology are fully understood and incorporated in the project's outline affordability appraisal. It should ask how robust the project's finances are, if for example, it were to be impacted by delays or increased resource requirements.

When considering the use of technology within a project, the degree of technological maturity being used in the project should be analysed. Is the technology well proven, or is the project considered to require an advanced level of technology, potentially even new, unproven technology? Can the required level of technology be delivered on time in order to meet the project schedule? Having completed the Technological Feasibility study, the project team should have a view as to whether or not the project is technically deliverable.

A thorough feasibility study should therefore be undertaken at the start of the project. In order to facilitate the study, an overview of the project goals detailing the required timeframe, available budget and required quality levels should be defined. A study of the challenges and risks faced in project execution should be made including the effect of any terms, conditions or other impacts affecting the project's delivery plan

Each project will be subject to certain legal requirements, both in terms of project outputs and the environment into which they are delivered. The legal section of the Feasibility report will review these and if successful provide proof that the project complies with the legal framework.

Operational feasibility considers the extent to which the project is suited to the requirements of the project sponsor, and to what extent the project is capable of delivering the proposed benefits in a way that they can be operated alongside, and integrated with, the organisation's existing systems. One key area to consider as part of an interoperability study is whether or not the new systems will cause dis-benefits to the organisation as a result of degradation of systems that are required to operate alongside the project output. Part of the issue in deciding this may be whether or not there is, within the organisation an ability to fully understand the end state derived post project delivery. The more complicated and ground-breaking the project, the more likely this problem becomes. This could lead to further question relating to the clarity of the environment in which the new system will operate.

The clarity of the environment in which the project is both constructed and delivered into can be impacted by cultural differences. Cultural differences can both enhance and inhibit the successful completion of projects.[5] The number of cross-cultural, international projects is increasing, but issues caused by cultural differences are all together common even between corporate organisations and even within the individual departments of these organisations. Many studies have been made of these differences and the way they influence outcomes.[6–8] The studies show that companies, groups or individuals may operate in certain ways that influence successful outcomes. Understanding them and anticipating the impact of them will therefore affect the chances of success.

Youker,[9] states that to be effective, the project manager 'must know what the values, beliefs and norms are in that culture'. These approaches will change over time, and the project manager should continually strive to understand them by all means possible, by researching the relevant culture, talking to people within the organisation in which the project is being deployed, and both listening and discussing the project implications with them. This will help align the values, beliefs and norms of the organisation and help align them with other cultures within the project teams.

Consideration of the project's safety is of such paramount importance that it could be argued to fall within the legal area of feasibility studies. It will be governed by a number of relevant laws and statutes which will have the power to impose significant penalties on projects that transgress. Construction projects which suffer fatalities during their lifecycles can be shut down for many months while government Health and Safety Executives carry out investigations. This will cost the project both time and money, but of course the impact is far deeper than this as the value of human life goes far beyond the impact on the project. In addition to safety, other key areas that need to be considered when reviewing feasibility, are shown in Figure 3.1 below.

In addition to the feasibility study areas outlined in Figure 3.1, sustainability is increasingly becoming an area of consideration within feasibility studies, driven in part by specific legal requirements.

3.2.1 *Sustainability in Projects*

The inclusion of sustainability goals, which are often interpreted as environmentally oriented, can cause a conflict between long-term environmental benefit and short-term economic operational goals. It has however been suggested that incorporating sustainable principles within, for example, construction projects can enhance the project's profitability. Sustainable practices such as reusing rubble from a brown field site and turning it into aggregate onsite reduces the costs of purchasing aggregate, transporting old rubble to landfill sites and removes the need to pay landfill taxes.

The aims of a sustainable approach are wider than just consideration of the environment. The 'triple bottom line' or 'people, profit and planet' concept of sustainability has been widely accepted[10] for a number of years, and puts forward the view that economic prosperity, social wellbeing, and the balanced use of natural resources, are all interrelated. Consideration of the triple bottom line has implications for all project management processes. For example, stakeholder management is tied to the people aspect and both considers and reflects the views and opinions of key stakeholders relating to sustainability issues. Budgeting will be impacted by the approach taken to sustainability driven costs and indeed even the degree to which future costs might be discounted in light of relevant views on sustainability issues. The approach to project risk management will also echo attitudes as to how the project's impact on planet eco-systems should be addressed. All these factors, and more, show how the approach to sustainability impacts the project's feasibility study, colouring the view of how achievable project success is through, for example, the economic or technological lenses. Similarly, the best time to begin considering a project's sustainability credentials is at the start of a project just as the feasibility study is undertaken. The link between a project's feasibility study and

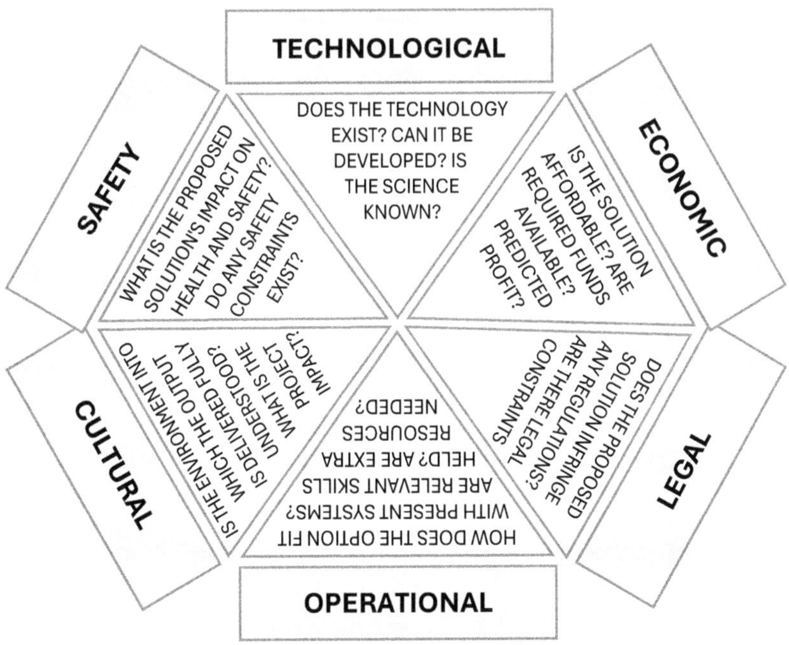

Figure 3.1 Feasibility study areas and outline considerations.

sustainability has been made[11] and the view established that in order to improve the way existing projects contribute towards sustainable development, all three dimensions of the triple bottom line must be addressed. The labels of economic, social and environmental issues can add value to the feasibility study by being used as headings under which to group the feasibility issues.

One approach that many organisations take throughout their operations and hence to their projects, is to adopt a CSR programme. Guidance is given in ISO 26000 which addresses how businesses and organisations can operate in a socially responsible way and lists the benefits of adopting a CSR programme as:

- Providing a competitive advantage.
- Enhancing the organisation's reputation.
- Increasing the ability to attract and retain workers, members, customers, clients and/or users.
- Helping to maintain employee morale, commitment and productivity.
- Improving the perception of stakeholders such as investors, owners, donors, sponsors, the financial community, governments, the media, suppliers, peers, customers and the community in which it operates.

ISO 26000 recommends that the organisation does this by developing a focus in seven key areas. These areas are, organisational governance, human rights, labour practices, the environment, fair operating practices, consumer issues and community involvement. These echo the triple bottom line of people, profit and planet mentioned above, but place more emphasis on people and planet rather than profit. The profit aspect need not though be divorced from the other two. There has traditionally been a take-make-use-dispose approach to resource consumption, which results in high levels of waste and the need to use land fill sites to dispose of this waste, all of which adds cost to any project. If the focus can instead be placed upon a repair, refurbish and reuse approach, then this has the potential to reduce costs for the project. This would mean that project outputs would need to be designed to be disassembled and reused, to be built to be more durable, and to move away from any trace of inbuilt obsolescence.

3.3 Project Review: The impact of Feasibility Studies on the Project: – The Garden Bridge Project

3.3.1 *The Project Background*

The idea of a "floating paradise" was originally suggested as far back as 1998. The resulting Garden Bridge project was a proposed new pedestrian bridge supporting a specially designed garden which was to span the River Thames in London from the top of Temple underground station across to the South Bank. It was to be funded both by public funds, funnelled through the Department of Transport (DoT) and using funds from Transport for London (TfL) with both providing £30 million towards the project.

Agreement to go ahead with the project and the delegation to TfL of the mayoral powers to 'develop and help enable the footbridge project' in order 'to promote social development and the environment in Greater London' was given on 27 August 2013.[12] TfL was tasked with establishing a clear policy statement detailing the need for a crossing of the Thames at this point, developing strategies for procurement, funding and for its delivery and construction. The project objectives went much wider than transport benefits such as those that TfL would normally focus on. It was seen as a public amenity, something that would not only increase foot traffic across the Thames at this point, but also attract extra visitors to the area and encourage interaction between the local economies on both sides of the bridge. Additional project benefits were seen as the enhancement of the quality of life, and an improvement in the safety and security for Londoners, through improved transport opportunities and a lower impact of transport to climate change as well as improving its resilience.[12]

As of 2016, the overall project was budgeted at £185 million with the remaining £125 million, in excess of the DoT and TfL funding, funded by private donations.

3.3.2 The Project Issue

A special trust, the Garden Bridge Trust was set up to oversee the design, build and maintenance of the bridge.[13] Early design work on the bridge was conducted in January 2013. Planning permission was granted in 2014 and construction of the bridge was due to start in September 2015. But construction of the project did not begin.

A review of the project, conducted by Dame Margaret Hodge[14] was commissioned by the new incoming Mayor of London in 2016. The review stated that decisions taken on the Garden Bridge had been driven by political electoral cycles and that value for money had been overlooked. From its inception, the report said, there had been confusion as to the purpose of the project. The funding of the bridge had become a major issue. Originally the cost had been put at £60m, and the project was to have been funded solely through private finance. It was deemed that the budget would need to increase still further amid continuing cost risks and uncertainties, and this at a time when two major donors had withdrawn and when doubts existed relating to the willingness of donors to contribute to a project that had, arguably, begun to lose public support.

The project had an unconvincing business case that was produced too late in the project lifecycle.[14] This was due in part to the perceived confusion surrounding the purpose of the project and hence its feasibility. Reference back to Figure 3.1 would lead to the questioning of the economic, technological and operational aspects of any feasibility study. The increasing costs show that the project was not feasible given the original budget. The technological aspect of a feasibility study, and discussion of how the garden would be funded and maintained once built, appear incomplete. The realisation of this resulted in the need for increased public, rather than private funds which contributed to the project's downfall.

3.3.3 How Could a Feasibility Study Have Helped?

Risks identified early on in the project[12] showed there was a possibility that suitable third-party funding for the construction, maintenance and operations of the bridge would not be secured and concluded that "in the event of this happening, the project will not proceed'. This would indeed prove to be the case. To mitigate this, in part it was stated that TfL would adopt 'sound project management techniques', something which Dame Margaret Hodge's report[14] suggested did not happen in full.

It is disappointing that having identified the risk relating to the failure to secure third-party funding, this risk was still allowed to occur. Taking just this one aspect of the feasibility study approach, more research could perhaps have been undertaken to quantify more accurately the amount of finance required and the likelihood of this being obtained. The lack of this research was not helped by reported confusion[14] over the purpose of the Garden Bridge. This deficiency, this lack of definition, hindered any attempt to judge the economic feasibility of the project, thus increasing the difficulty in securing private finance.

These issues were recognised in part by the project, but the project then failed to adequately explore them and failed to put it place sufficient mitigation measures. A fully developed feasibility report would have identified and quantified these issues earlier and given the project a chance to reduce the £53.3m of costs reported in a TfL press release[15] to have been spent prior to the cancellation of the project in August 2017.

3.3.4 *Conclusions on the Garden Bridge Project*

There were many issues with the project, such as an unclear vision, an unsound business case, developed later than it should have been, and a lack of project governance and financing problems.

All these issues could have been alleviated by the development of an early, detailed feasibility study, forcing project attention and decision making around key aspects of the purpose and deliverability of the project goals. Even if these goals were to be changed at a later date (through a change control process – see Chapter 12) a thorough feasibility study would have vastly improved the understanding of the project and its chances of success.

3.4 The Feasibility and Sustainability Before the Birth of Modern Project Management*

(* In this section, the research relating to the construction of HMS Barham 1912–1915 mentioned in the introduction is drawn upon to help illustrate how the subject of this chapter has developed.)

Archival research relating to project methods prior to the birth of Modern Project Management has not revealed any sustainability considerations or feasibility study in the construction of the HMS Barham case study.

Dealing with sustainability first, this is not surprising given the relative newness of sustainability as a concept, it was not included in the APM Body of Knowledge as a subject in its own right until BoK6 in 2012. During the construction of HMS Barham, it would have been survivability that was a concern rather than sustainability!

The approach to constructing warships was relatively unique from a project and programme perspective. Each year a budget for approximately four or five ships was approved by Parliament. This was termed, in the case of HMS Barham, the '1912 Shipbuilding Programme'. Hansard records[16] that as part of the debate over the 1912 programme, the question of how many Dreadnoughts should be built was debated before the Naval Estimate was passed. The programme of ships evolved over the years, and even between ships within the programme, making each vessel unique to a greater or lesser extent. It is quite possible therefore that an individual ship would not have had a feasibility study, or equivalent, undertaken. Feasibility would have been assessed at a programme level, particularly in respect of economic, technological and operational issues, and each year's shipbuilding programme would have shown incremental improvement on the year before.

What is clear though, is that the economic feasibility of the vessel was judged through the estimating process (see Chapter 7) and subsequent parliamentary approval of programme costs.

References

1 Project Management Institute. (2021). *The standard for project management and a guide to the project management body of knowledge* (7th ed.). Project Management Institute, Inc.
2 APM. (2019). *APM body of knowledge* (7th ed.). Association for Project Management.
3 Ohara, S. (2017). *P2M – A guidebook of project & program management for enterprise innovation* (3rd ed.). Project Management Association of Japan (PMAJ).
4 Naughton, E., Education, I.P.M.A., & Barnes, M. (2013). The iron triangle under threat. *PM Word Journal,* 2(12), 1–4.
5 Anbari, F. T., Khilkhanova, E. V., Romanova, M. V., Ruggia, M., Tsay, C. H.-H., & Umpleby, S. A. (2010). Cultural differences in projects. Paper presented at PMI® Research Conference: Defining the Future of Project Management, Washington, DC. Project Management Institute. https://www.pmi.org/learning/library/differences-projects-multicultural-culture-leadership-6478 [Accessed 17 October 2024].
6 Hofstede, G. (2000). *Culture 's consequences: Comparing values, behaviors, institutions, and organizations across nation* (2nd ed.). Sage Publications.
7 Aycan, Z., Kanungo, R. N., Mendonca, M., Yu, K., Deller, J., Stahl, G., & Khursid, A. (2000). Impact of culture on human resource management practices: A ten-country comparison. *Applied Psychology: An International Review, 49*(1), 192–220.
8 Trompenaars, F., & Hampden-Turner, C. (1998). *Riding the waves of culture: Understanding cultural diversity in global business* (2nd ed.). McGraw Hill.
9 Youker, R. (2004, June). What is culture in organizations? *Proceedings of the International Project Management Association 18th World Congress on Project Management* [CD]. Budapest, Hungary.
10 Elkington, J. (1997). *Cannibals with forks: The triple bottom line of 21st century business.* Capstone Publishing Ltc.
11 Shen, L.Y., Tam, V.W., Tam, L., & Ji, Y.B. (2010). Project feasibility study: The key to successful implementation of sustainable and socially responsible construction management practice. *Journal of Cleaner Production, 18*(3), 254–259.
12 Mayor of London, London Assembly. (2013). MD1248 temple to south bank footbridge development proposals. https://www.london.gov.uk/decisions/md1248-temple-south-bank-footbridge-development-proposals#_blank [Accessed 17 October 2024].
13 The Comptroller and Auditor General. (2016, October). *Investigation: The department for transport's funding of the garden bridge.* https://www.nao.org.uk/wp-content/uploads/2016/10/Investigation-the-Department-for-Transports-funding-of-the-Garden-Bridge.pdf [Accessed 17 October 2024].
14 Mayor of London, London Assembly. (2017). Independent review of the garden bridge project. https://www.london.gov.uk/independent-review-garden-bridge-project [Accessed 17 October 2024].
15 Transport for London. (2019). Transport for London publishes detailed final cost of the undelivered garden bridge project. https://tfl.gov.uk/info-for/media/press-releases/2019/february/transport-for-london-publishes-detailed-final-cost-of-the-undelivered-garden-bridge-project [Accessed 14 October 2024].
16 Hansard HC Deb. (1912, May 16). vol 38 cc1271–2. https://api.parliament.uk/historic-hansard/commons/1912/may/16/british-shipbuilding-programme [Accessed 24 October 2024].

Chapter 4

The Project Lifecycle

A project lifecycle is something that is shared by all projects, yet although having a common basic form it can vary from industry to industry and indeed from company to company and even between projects within the same organisation.

Once agreed the lifecycle of the project becomes a 'living' document having the potential to evolve throughout the duration of the project.

As a living document it will need to be kept under constant review by the project manager in order to reflect changes in the project's definition or progress caused by both internal and external changes.

4.1 The Purpose of the project lifecycle

The project lifecycle is described by the APM as providing the 'set of distinct high-level stages required to transform an idea of concept into reality in an orderly and efficient manner'.[1] The PMI[2] define the Project Life Cycle in a similar way, referring to the series of phases that a project passes through as it delivers the final output required. The PMI also highlight the contribution of the Project Life Cycle in optimising the project outcomes in terms of business benefit and other stakeholder value throughout the project. The type of Project Life Cycle is determined by the approach to the delivery of the project outputs, or outcomes, and the 'rhythm' of that delivery. The rhythm of delivery is important since it defines when, within the life cycle, project outputs will be available. The rhythm of delivery therefore directly impacts on the timing and realisation of the benefits discussed in Chapter 2. This in turn, will influence the number of phases or sections within the life cycle. These phases are discussed in more detail within the sections below.

Both the APM and PMI make the distinction between what is termed the project lifecycle and the extended project lifecycle. Which one is appropriate for any particular project is dependent on where and when the project comes to a finish, or alternatively, which phase the lifecycle concludes with. In the case of the Project Life Cycle this will happen once the outputs of the project are operational. It will not include the operational phase itself as this is the domain of the 'extended project lifecycle' which will continue to the end of the products life and also include the eventual end of use or disposal phase. Where the lifecycle includes the requirement to support the output throughout its time in use, this may involve activities such as product support and the ongoing adaption to market requirements. This 'extended project lifecycle' is often termed 'product management' since due to the lengthy time that an item might be in use or available within the market, the focus of management shifts to the product rather than the project that produced it.

DOI: 10.1201/9781003470748-5

The basic purpose of the project lifecycle outlined by the APM at the start of this section, is also reflected in the P2M documentation,[3] where it states that a project lifecycle 'consists of beginning, middle, and end phases' which can be subdivided. Whilst this is not particularly detailed, it illustrates that there must be a progression and logical order of activities through the project lifecycle, meaning that the lifecycle gives form and structure to a project, helping to ensure its success. The CIOB[4] agree with the sentiments of the individual project management bodies above, and consider the asset itself to have its own lifecycle extending from cradle to grave. It recognises that the project team itself may not be involved in all these stages of the asset's lifecycle drawing the same inference as the APM and PMI, in recognising the difference between the project lifecycle and the extended project lifecycle. The CIOB however do make an interesting observation in that regardless of how many of the asset lifecycle stages the project team is involved in, the client or project owner will have ultimate responsibility across the complete asset lifecycle.

4.2 How Is the concept of Project Lifecycle Applied Within Projects?

The earlier part of this chapter established that the project lifecycle's purpose is to provide form and to outline progression within the structure of a project.

If the application of a project lifecycle is to be discussed however, then it is necessary to understand what is meant by 'a project' and where it sits within the raft of activity undertaken by many of today's firms. The APM[1] define a project as a 'unique transient endeavour undertaken to achieve planned objectives'. The PMI[2] support this and define a project similarly as a 'temporary endeavour undertaken to create a unique product, service, or result.' A project therefore delivers a unique output designed to meet an organisation's planned requirements. One project though can't necessarily deliver all the change required by an organisation and so programmes, sometimes defined as 'a group of projects' have evolved. Programmes however are more structured than just a group of projects. They will be 'selected or commissioned, planned and managed in a co-ordinated way' so as to deliver multiple objectives, and in doing so meeting multiple business objectives.[5] This contrast between projects and programmes and the multiple nature of projects within a programme, leads to Table 4.1 below which summarises some of the main differences between the two.

Table 4.1 serves to illustrate clearly some of the specific qualities attributable to project delivery. How then is the project delivered? Through which steps or stages must it pass?

There is no standard project lifecycle that is right for all industries. The lifecycle will vary from business sector to business sector. Each lifecycle though will follow a similar basic pattern:

Each one of the phases in the lifecycle will consist of processes such as planning/ controlling/ executing and closing. These can be repeated as often as necessary within each phase, and this highlights one of the main differences between project phases and project processes. Phases have a beginning and an end. Processes, however, are continuous and occur throughout the lifecycle. Planning for example will be part of each project phase. There is no 'Planning Phase' as such, where all the planning will be done. There may be phases where more planning is done than there is in other phases, but planning will be a part of every phase.

The purpose of any project's phases is to drive forward and enable the delivery of strategy into operational being or, into the operational environment. This will allow company policy and strategy to become reality and to shape the organisation's sphere of operations. As shown in

Table 4.1 Characteristics of Projects and Programmes

	Characteristics of	
	Project Delivery	*Programme Delivery*
Nature of Delivery	A project involves intense activity focused on delivering predetermined outputs	A broad activity or set of activities focused on delivering business change realising a wider set of benefits than one project on it's own can
Scope Definition	A project is suited to well scoped and bounded outputs that can be well defined	A programme is suited to complex activities delivered within a dynamic and uncertain environment with fluctuating interactions and relationships
Benefit Realisation	Benefits within a project are delivered at the end of the project once the project outputs have been delivered	Benefits can be delivered after the programme finishes but also during the programme as outputs from contituent projects are put into service
Duration	A project has a predetermined start and finish date	Unless deemed to have become invalid and hence to have been closed early, the programme will usually continue until the desired end state has been achieved. This could require additional projects to be undertaken

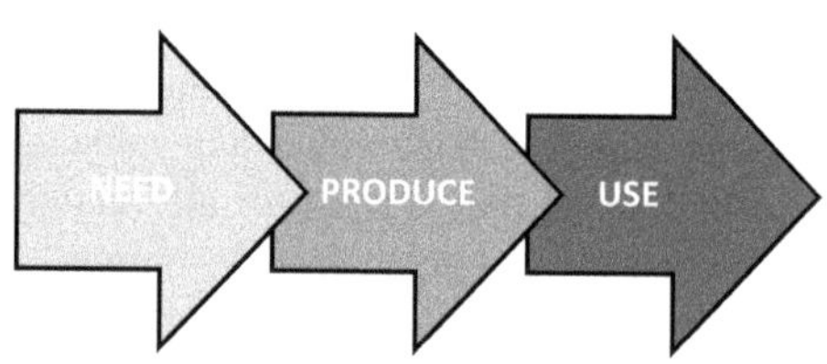

Figure 4.1 A basic project lifecycle.

Figure 4.1, the project recognises a need and then initiates a set of actions to produce an output at which point it may hand this output over to the end user or it may continue to manage the outputs through the in use and disposal phases. As stated above, if it does not include these last two phases, the project has what is referred to as a 'project lifecycle'. If the project does include these phases, then it has what is termed an 'extended project lifecycle'.

4.2.1 *The Benefits of Using a Project Lifecycle*

The use of a project lifecycle has a number of benefits. Apart from providing a logical progression for the project activities, it provides natural break points (see approval gates in Figure 4.3) or places where the continued viability of an ongoing project can be validated. If a project is no longer viable it can, of course, be terminated thereby preventing any additional, wasteful spending of time and money. At these breakpoints a project review will be undertaken, which serves to force approval of the project's ability and suitability to progress into the next phase of the lifecycle.

A lifecycle will also give structure to a project, indicating when certain roles are needed, for example enabling forward planning of engagement with testers, end users, or any of the project's other stakeholders at specific times within the project duration.

If all project lifecycles provide these benefits, but as has already been noted, there is no one standard project lifecycle, how are lifecycles structured and what benefits do the different forms of lifecycle offer?

There are a number of different forms of project lifecycle each of which can help the project manager deal with a variety of different factors presented by projects. Perhaps the most well-known project lifecycle is called the 'waterfall' lifecycle. Various different bodies will each have their own variant of the waterfall project lifecycle, each will however have a similar form to that shown in Figure 4.2.

The diagram clearly shows the difference between the project lifecycle and the extended project lifecycle, but it is not immediately obvious from the diagram why it is called a waterfall lifecycle.

The phases in a waterfall lifecycle are very structured and the project will flow from one to the next without returning to an earlier phase. This is perhaps shown more clearly if the lifecycle is drawn as follows in Figure 4.3.

Here, the reasoning behind the waterfall name is clear – the project is designed to flow from one phase to the next in an orderly fashion without reverting to an earlier phase. As the project passes from one phase to the next, there is a checkpoint, or Approval Gate at which point the project's progress to date will be assessed along with the achievability of its end goals, its continuing strategic alignment and its readiness to move onto the next phase. If the review of all these factors is positive, then the budget for the next project phase will be awarded to the project and it will progress into the next phase.

At each Approval Gate the products of each phase will be reviewed and entered into the project library. There are some key documents that all projects need to provide in order to run

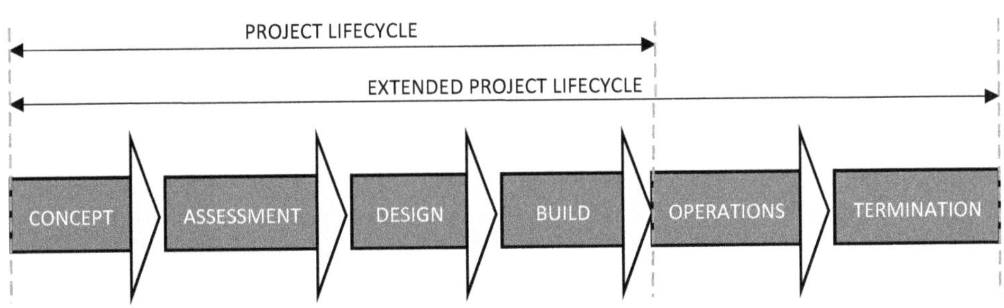

Figure 4.2 Project waterfall lifecycle.

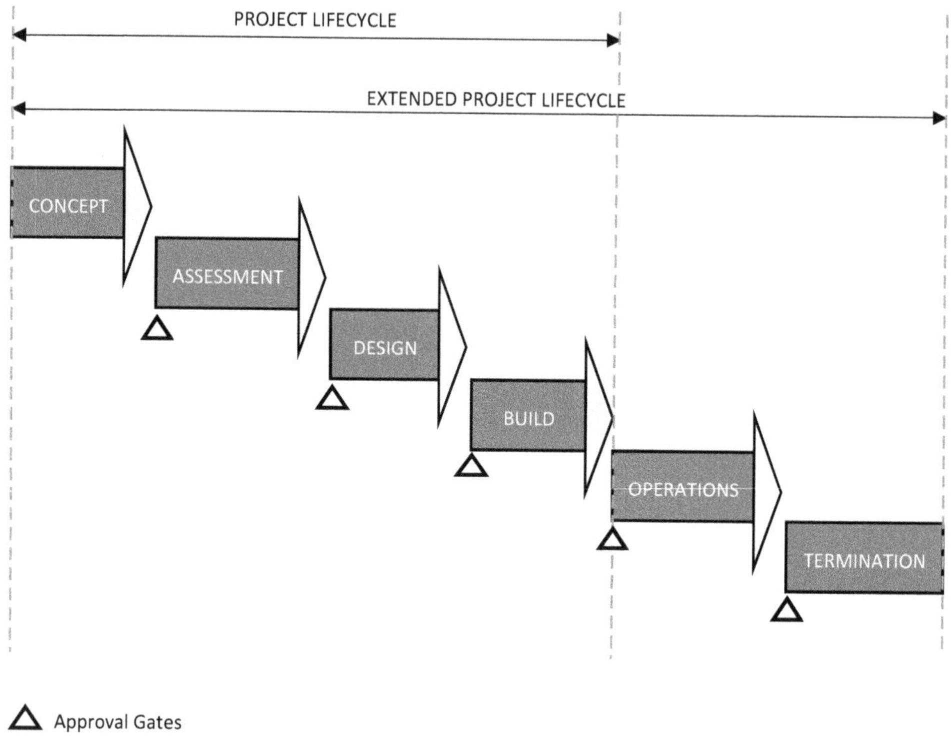

Figure 4.3 Project waterfall lifecycle with approval gates.

efficiently and effectively. Many of these are discussed elsewhere in the text, but an overview of these is provided below in Figure 4.4, which also shows their indicative position within the lifecycle and the development in project definition that is implied by the timeline as the project advances through the concept to termination phase.

One of the key documents pictured in Figure 4.4 which is not mentioned elsewhere, is the Project Management Plan which describes the environment in which the project will be managed. Key documents such as the project's communication strategy/ health and safety plan, as well as the risk management plan will be contained in the plan and will govern the way in which the project is delivered throughout its lifecycle. Table 4.2 below contains a list of typical contents for a Project Management Plan.

The waterfall project lifecycle (shown in Figure 4.3) adds structure and a clear organisation by which the project can reach its objectives. It can be used a simple planning tool. Top level staffing requirements can be attributed to the key activities in each project phase together with an outline of other resource requirements and their lead times.

If a waterfall lifecycle has a weakness, it is that it implies a straightforward progression through the project delivery phases. To enable such a smooth progression through the phases, both the project outcome and the environment into which it delivers, must be clear and certain, meaning that the project objectives must be well defined and stable. If this is not the case, as it often isn't, then this should be reflected within the project's proposed lifecycle. The simplest change that could be made to the waterfall lifecycle model, would be to modify it by allowing the preceding phases to be revisited if necessary, or even overlapped as shown in Figure 4.5.

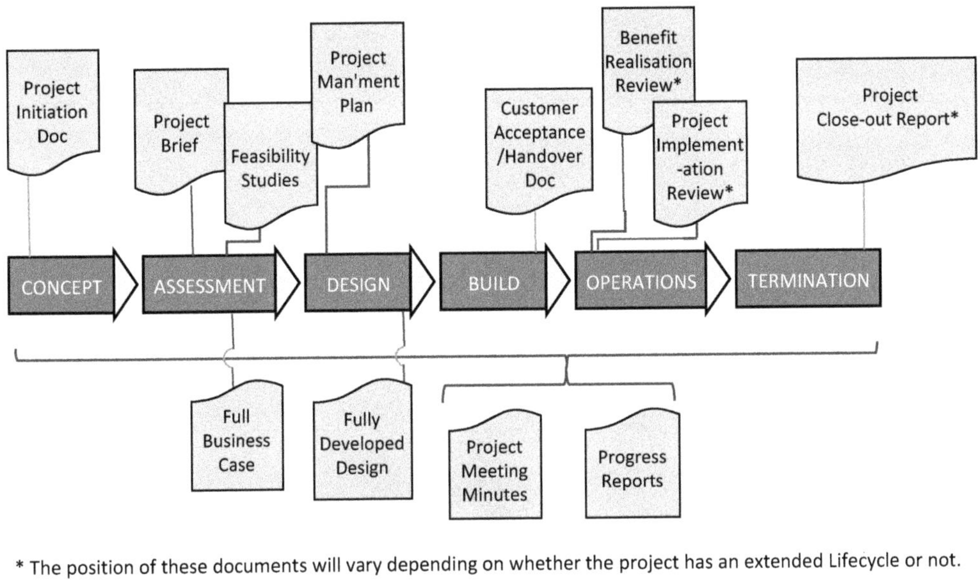

Figure 4.4 An example of project documentation.

Table 4.2 Typical Contents of a Project Management Plan

Introduction & executive summary
Project team members
Project organisation, roles & responsibilities
Business case, budget, commitment acceptance
Project deliverables, scope, work breakdown structure, top level schedule
Risk, changes, quality, H&S communication project management procedures/plans
Critical assumptions and constraints

The modified waterfall lifecycle allows for a degree of exploration, or parallel running of project phases. In a project where the nature of the project deliverables and the nature of the post-project environment are well known, this adds time and cost to the project.

An Agile Lifecycle takes this approach one step further. It includes several iterations of product development so that it can develop the project deliverables in incremental steps. It is often seen in IT projects as it promotes adaptability, enabling a series of 'do-loops' meaning that the product development need not follow a pure linear path. This approach is shown in Figure 4.6.

The lifecycle in Figure 4.6 shows the multiple analysis/design/code and test cycles often seen in an IT development project or similar. The number of iterations of this part of the lifecycle which the project contains will vary from project to project depending on exactly what is needed.

Where there is some uncertainty within the project, both the Modified Waterfall and Agile lifecycles allow progression to the end state to be made more gradually and allow the goals to be refined as the project progresses.

The key question though, in the case of the lifecycles shown above, is how much uncertainty can be coped with?

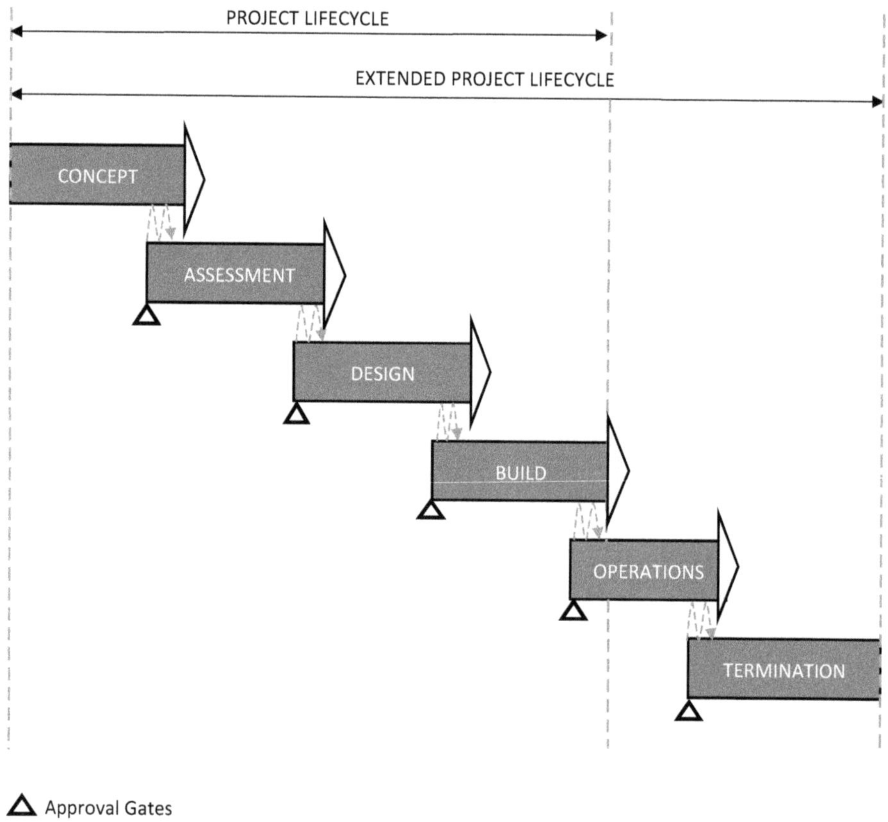

Figure 4.5 A modified waterfall lifecycle.

4.2.2 *Sources of Project Uncertainty and Their Impact on Lifecycle Choice*

A project will potentially have two major sources of uncertainty, namely, the type of the output required and the nature of the environment in which the output needs to operate.

To cope with the potentially large degrees of uncertainty in these two variables, alternative forms of lifecycles to the waterfall approach should be considered. It is important to note that they still follow the 'need – produce – use' format shown in Figure 4.1, but the way in which they deliver this can be quite different in appearance.

The first uncertainty factor to consider is when the environment into which the project output is being delivered is well known, but the exact characteristics and nature of that output are not so well known. A hypothetical example of this might be the future introduction of hover-cars into the current road transport. The current parameters of transport are well known, but how a hover-car would need to be developed in order for it to interact with the current environment is much less clear.

An Incremental Project Lifecycle enables these uncertainties to be addressed and allows the project's end product to be delivered through a series of stages. Figure 4.7 below shows how this approach to a lifecycle would look.

Each increment can be used to evaluate its effectiveness, whilst at the same time it can inform the development of the next increment and then the two can be employed together to deliver the

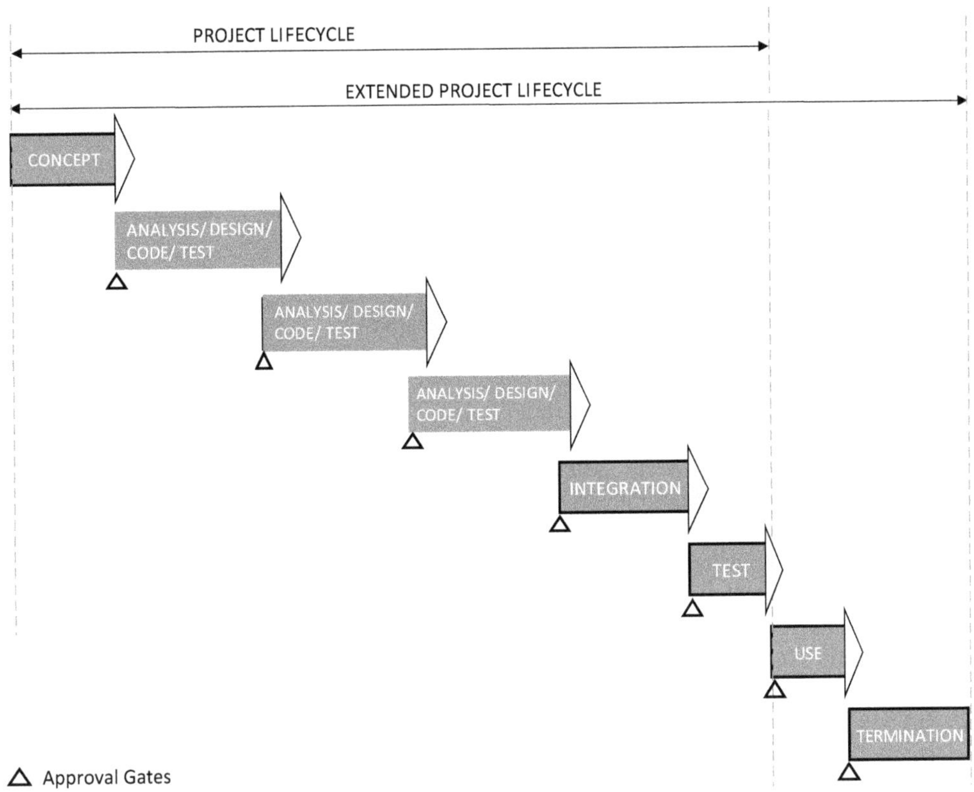

Figure 4.6 An agile or iterative lifecycle.

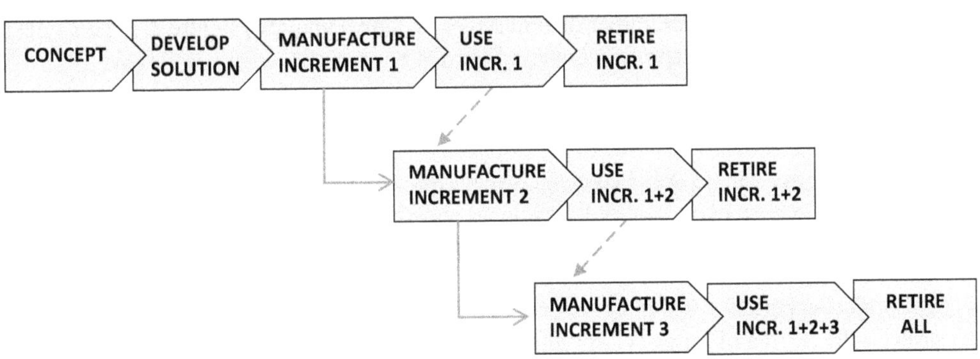

Figure 4.7 An incremental lifecycle.

project output. This can be repeated multiple times until the project lifecycle is complete and the outputs are all retired.

If the project is subject to greater uncertainties, where for example both the delivery environment and the product characteristics are not fully known, then a more extreme version of the project lifecycle would be required. Suppose a project manager was tasked with delivering a scientific observation outpost to be sited in the Mariana Trench in the Pacific Ocean, 35,814

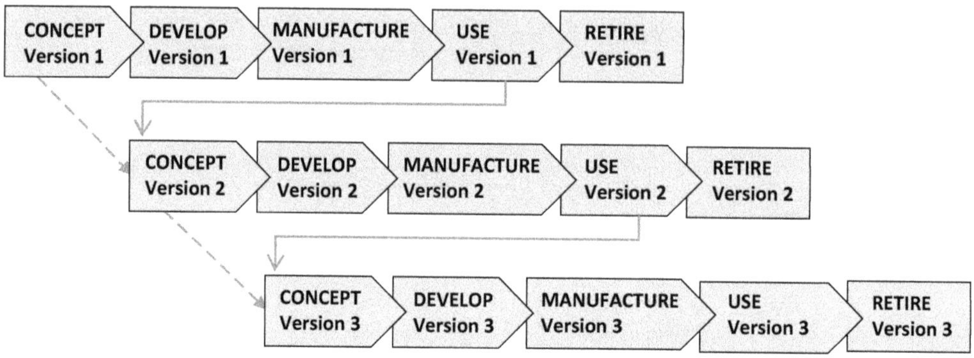

Figure 4.8 An evolutionary lifecycle.

feet, or over six and a half miles, below sea level. At this, the lowest known point on earth, little is fully understood about the environment at that depth, nor about the long-term effects of operating at such pressure. The exact requirements of the deliverable in terms of the tasks it would be required to perform could well be the subject of debate too, meaning that the product characteristics were still to be fully developed.

The resulting project lifecycle would need to be able to facilitate the dynamic changes required by the developing understanding of both the product and environment characteristics. Rather than build on an earlier product release, as shown by the lifecycle in Figure 4.7, a more revolutionary approach is required, where the first iteration of the product is deployed, lessons are learnt from it, and the characteristics become better understood meaning that a new and improved stand-alone version of the product can be developed. The lessons learnt from using the first iteration are embedded in the development of the second version and once Version 2 is in service, Version 1 is retired. The degree of change between the versions of the product can therefore be truly evolutionary, rather than iterative as shown above in Figure 4.7. This type of Evolutionary Lifecycle is illustrated above in Figure 4.8.

4.2.3 A Spectrum of Lifecycles

The lifecycles above can be sketched along a spectrum which illustrates their flexibility, from a waterfall lifecycle to an evolutionary lifecycle, as shown in Figure 4.9 below.

Each variant of the lifecycle not only has an impact on the way in which the delivery of project outputs is shaped but can also offer different benefits depending on variables such as the level of project experience within the project team, the dominant type of risk at the forefront of the project, the degree of customer certainty and the stability of the customer's requirements, and even the geographic distribution of the project team. Understanding these factors can help in the selection of the best lifecycle for a particular project.

The potential strengths and weaknesses of each alternative project lifecycle are outlined in Figure 4.10 below.

These potential strengths and weaknesses mean that in certain project scenarios, one type of lifecycle will offer greater benefits to the delivery of the project than another.

If, for instance, the main source of project risk is thought to be financial, a waterfall lifecycle may be preferred as it is easier to both estimate and monitor the project's cashflow and out-turn costs. If on the other hand, the major risk driving the project is a technical one, then an iterative

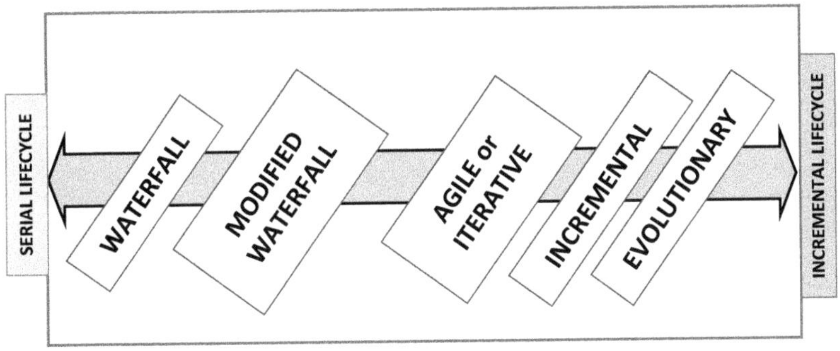

Figure 4.9 A spectrum of lifecycles.

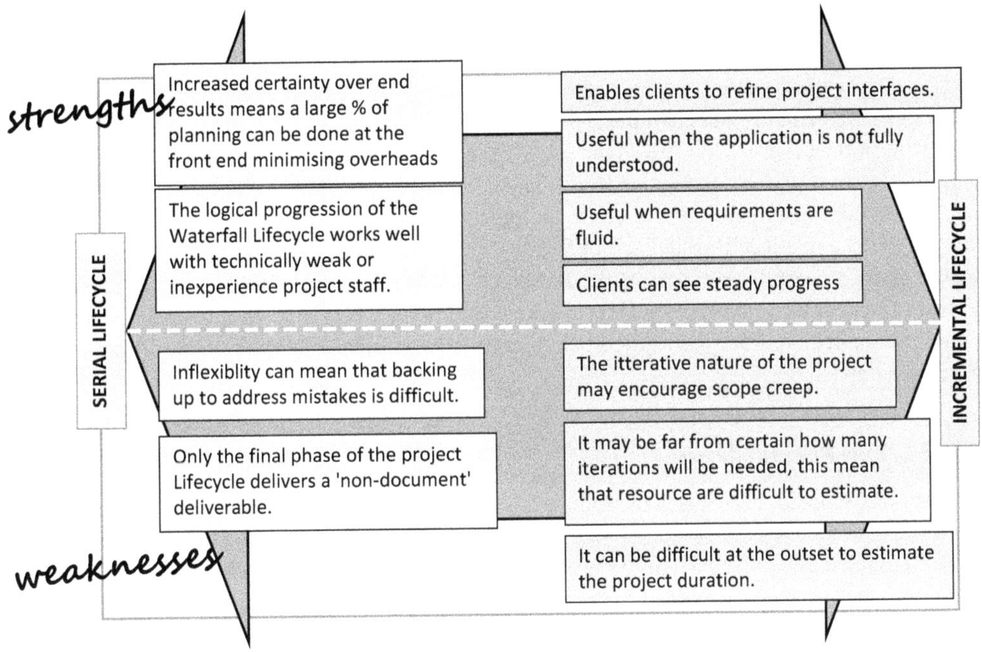

Figure 4.10 Project lifecycles strengths and weaknesses.

approach with increased experimentation and testing loops might be favoured. If the risk is linked to time, then an iterative approach could be argued to be best where some of the iterations or builds, have the potential to be set live early.

If the projects requirements are known, clearly outlined, and stable then a waterfall lifecycle could prove more suitable than an incremental approach. The reverse situation would of course favour the consideration of a more iterative approach. If the project team is stable, fully engaged in the project, co-located and experienced in project work, then an iterative approach could be suitable. On the other hand, if any of these four factors are not present within the project team, a more serial approach to the lifecycle should be considered as it will be easier to maintain project focus and ensure consistent project communication and reliable monitoring of project progress.

The points raised above, in particular those linked with project risk types, are often closely defined by the client requirements, and formulated in the project's time, cost and quality criteria.

The following section discusses a project where a mis-understanding of the lifecycle selection criteria led to a massive cost overrun for the project.

4.3 Project Review: The impact of the Lifecycle on the Project: Denver International Airport

4.3.1 *The Project Background*

Within project management there are a number of classic case studies of project failure, and the construction of Denver International Airport (DIA) certainly makes a case for being included in this body, with a reported cost overrun of more than $3. 3 billion and a delivery 16 months behind schedule.[6]

The DIA project was at the time of construction, the largest airport in the world. The complex in its entirety was twice the size of Manhattan, covering some 53 square miles, (13,727 ha), it had 5 separate main runways each approaching 2.5 miles, (3,660 m), long, enabling up to 1,750 planes to take off and land each day.[7] At the time of completion, DIA was, unsurprisingly, the largest airport in the world. The sheer size of the airport meant that the design of the baggage handling system was challenging.[8] It was usual in airport projects within Americas at the time to give carriers responsibility for construction of the baggage handling system. After initially doing this, as a result of the site's complexities it was decided that the baggage handling system would be built centrally by DIA.[9] Built to replace Stapleton Airport, DIA was designed to enable airlines to use it as a hub from which to connect to other airports, both local and international, which was a decision that multiplied considerably the number of bags passing through the airport each day.

4.3.2 *The Project Issue*

As previously noted, the causes of project failure are seldom singular, but some elements stand out more than others. One particular element of the DIA project, the automated baggage system is often highlighted[9,10] as a cause of the project's cost and time overruns. Within the delivery of the baggage system, the application and choice of project lifecycle certainly impacted the DIA project.

The scale of the airport and the distance between some of its gates, coupled with the requirement for bags to be moved quickly from one destination to another, sometimes up to a mile away, added considerably to the project's problems. The baggage handling system was not originally included in the project specification and its later inclusion was at the root of key project difficulties. Some writers[8] conclude that key mistakes were made when the complexity of the baggage handling system was initially estimated, and consequently insufficient time was allowed to properly develop the system. One reason that the estimates made might have been inaccurate, and hence the complexity underestimated, was that it had been 20 years since a similar airport had been built in the US – Dallas/ Ft Worth International. In this time the number of plane departures of US carriers increased from 4,777,000 to 8,426,000 per annum (a 76% increase), and perhaps more importantly, the number of passengers carried by US Airlines per annum increased from 209,092,000 to 559,638,000[11] (an increase in excess of 167%).

The development timeline, or lifecycle, for the baggage handling system element of the project was adversely impacted by the review, 2 years into the project, of the original decision that each individual airline would build its own baggage handling system. Initially it was considered that one airport-wide baggage handling system would add efficiency to the overall project, reducing the time taken to route the baggage within the terminal buildings and it was also thought to be an approach that would save costs. The delivery of the new plan, post the review, has been reported as underestimating the degree of complexity of the baggage system which needed to be expanded. It was reported too that the revision did not take into account the potential impact, or risks, of the newness of the proposed technology, nor of the high level of coordination needed. As a result, these challenges were not fully reflected in the project's system implementation plans,[9] meaning that the baggage handling system was approached and managed in the same way as the project overall, which was a large but fairly conventional construction project where a standard waterfall lifecycle approach was used.

The airport opened 16 months behind schedule in 1995. As late as 1994 it was realised that the plans for the delivery of the airport's baggage system were in need of restructuring. In September 1994, a deal to divide the baggage system into two separate systems[9] was agreed. When the airport did open in February of 1995, the original planned automatic baggage system was far from the fully automated system originally planned and instead it consisted of a semi-automatic baggage system serving United Airline's concourse whilst a manual approach was provided for the remaining concourses.[9]

4.3.3 *The Potential Impact of an Alternative Project Lifecycle on the Project*

The approach to the delivery of the baggage system was clearly beset by a number of issues, but the following quote highlights one in particular. In describing the first test of the baggage system, it has been said that the project 'invited reporters to observe the test. 7,000 bags were to be moved to Continental's concourse A to United's concourse B. So many problems were discovered that testing had to be halted. Reporters saw piles of discarded clothes and other personal items lying beneath the telecar's tracks'.[9] This quote causes a number of questions spring to mind. Was this really the 'first test' and if so, why? If it wasn't, then why was it so disastrous particularly given it was in such a public forum? Would a more iterative approach to development, delivered through and managed by a more Incremental Project Lifecycle have been advisable? Should what was known to be a complicated part of the airport development have been treated differently to the rest of the construction project which could be seen as a large but not unusual construction project?

If the baggage system had used a more iterative approach and gone through more rounds of testing and improvements, as laid out by an iterative or agile project lifecycle, potentially the public humiliation of the 'first test' could have been avoided. If this approach had been extended back to the start of the project, potentially some of the issues could have been identified earlier and project delays, calculated as costing more than $1m a day, would have been avoided.

4.3.4 *Conclusions on DIA and the Lifecycle Chosen*

Failure to recognise risks is given as a reason behind the late delivery and the final budget being three and a half times the original budget (it increased from $1.7bn[1] to over $5bn[6]). As discussed, mitigation of the risks associated with the baggage handling system could have been addressed by taking advantage of the increased number of design and testing cycles offered by an iterative lifecycle approach. This might have been considered too time consuming at the outset, but its use could well have resulted in less delay overall. The big bang approach illustrated by the quote describing the 'first test', served to introduce far more risk to the project and is indicative of the baggage system being considered simple and straight forward despite its size. The baggage handling system was the first to be built in an American airport for 20 years and as such the requirements of the system, due in part to the volume of bags being catered for and also to the size of the airport, were relative unknowns. Where the characteristics of both a system being delivered and the environment into which it is delivered are unknown, an iterative project lifecycle approach enables greater flexibility and enables the project to take advantage of learnings from each iteration thereby increasing the chances of project success.

4.4 The 'Project Lifecycle' Before the Birth of Modern Project Management*

(* In this section, the research relating to the construction of HMS Barham 1912–1915 mentioned in the introduction is drawn upon to help illustrate how the subject of this chapter has developed.)

Well before the construction of DIA and prior to the formalisation of Modern Project Management, the evidence shows that within the construction of HMS Barham (1915), the word 'project' does not appear in the records, and as a result, neither does the phrase 'project lifecycle'. Perhaps surprisingly though, the word 'programme' does appear. Each year, in this period, the Admiralty costed and approved the number of warships to be built. Typically, there would be a group of four or five vessels commissioned each year.

HMS Barham was part of the 1912–1913 programme consisting of HMS Queen Elizabeth, Warspite, Valiant, Barham and Malayia. Each subsequent design was based on the first in class, in this case HMS Queen Elizabeth. The class itself was largely a development of the design of the previous year's shipbuilding programme. Parts were designed and shared across the class, clearly demonstrating features of today's APM definition of a programme,[1] which states that a programme incorporates a number of related projects.

Within the Admiralty shipbuilding programme of 1912–1913, documentary evidence found for a single document akin to a project lifecycle is slim. The unanswered question of course being, is the comparative absence of records due to the lifecycle records not being preserved or due to the concept of a lifecycle being, at the time, undeveloped?

Components of a standard lifecycle though are present and there is evidence that dates such as delivery points could be predicted, which implies organisation and the existence of the predictive power of a lifecycle or similar document. The document below dated 18/9/15 shows the expected delivery dates of future ships with HMS Barham shown as October 1915 as highlighted.

The document predicts delivery dates up until December 1916, more than a year after the document was written in September 1915. At the time, this information would have been very important in order for the Admiralty to plan the replenishment of any vessels lost in the war in order to ensure its ability to maintain naval superiority.

To achieve this, the Admiralty would have needed a systematic approach to the construction and delivery of every ship.

This systematic approach is evidenced by the author's own diagram below at Figure 4.11, which shows the timing of various build activities within the construction of HMS Barham.

Study of Figure 4.11 below shows that the order in which the tasks are recorded fits very much with the expectations of a similar project today, following a natural progression from

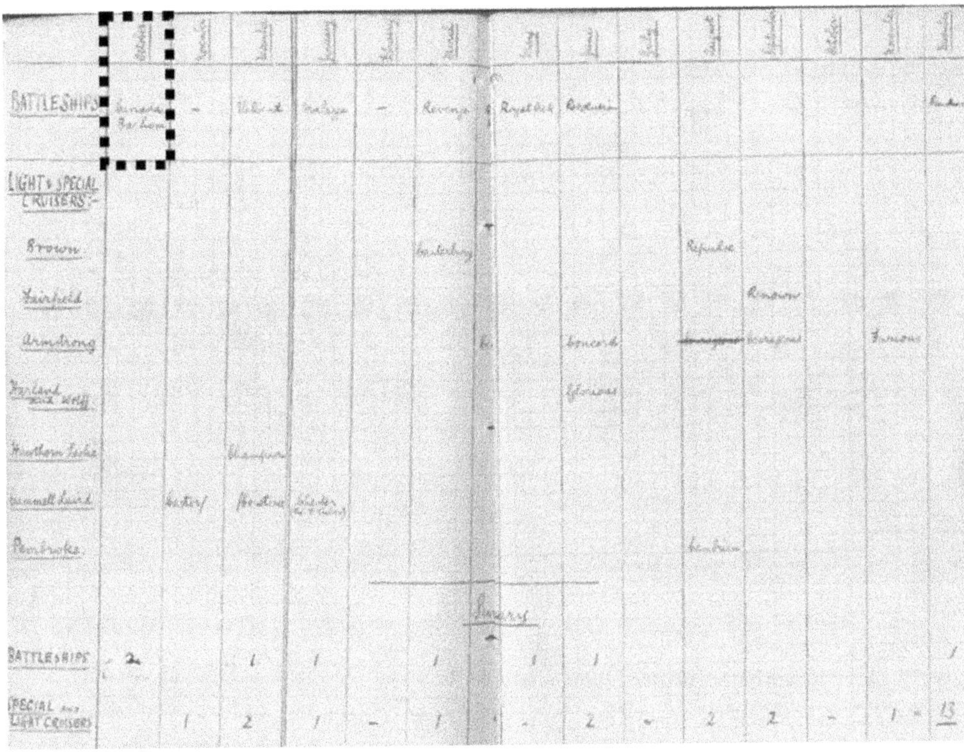

Photo 4.1 Projected delivery dates as at 18/9/1915 for HMS Barham. By permission of The National Archives of the UK: ADM 1/8435/297.

planning activities, letting contracts, through construction to acceptance testing and ultimately to customer acceptance. The references to the labour required by HMS Barham indicate that there was a process of monitoring and control through the lifecycle enabling both build progress and labour shortages to be identified and acted upon. The 1909 Instructions to the Professional Officers of HM Shipyards (see Photo 5.1) found in the National Archives was inscribed with the date 1915 suggesting it was current at the time of HMS Barham's construction. At the time of HMS Barham's build, the John Brown Yard was Admiralty controlled,[12] and so would have been impacted by these 1909 Instructions, indicating therefore that there was an organisational design established at the start of the project, as would be expected within a Modern Project Management lifecycle.

Today's APM's project lifecycle[1] shows a project to have four main phases.

Each of the documents shown in Figure 4.11 below would fit sequentially into the phases shown in the APM's lifecycle (Figure 4.12). The latest document shown in Figure 4.11 is the as built acceptance from the Admiralty showing that the project, in today's project management terminology, had a project lifecycle rather than an extended one. Given the nature of the asset this is of course what would be expected, to issue a contract at the time of the First World War to both construct and operate a warship would have been unthinkable and today to operate such a project output clearly still requires highly trained skilled personnel within our armed forces rather than staffing provided by the manufacturer, hence a hand-over upon completion would be exactly as expected.

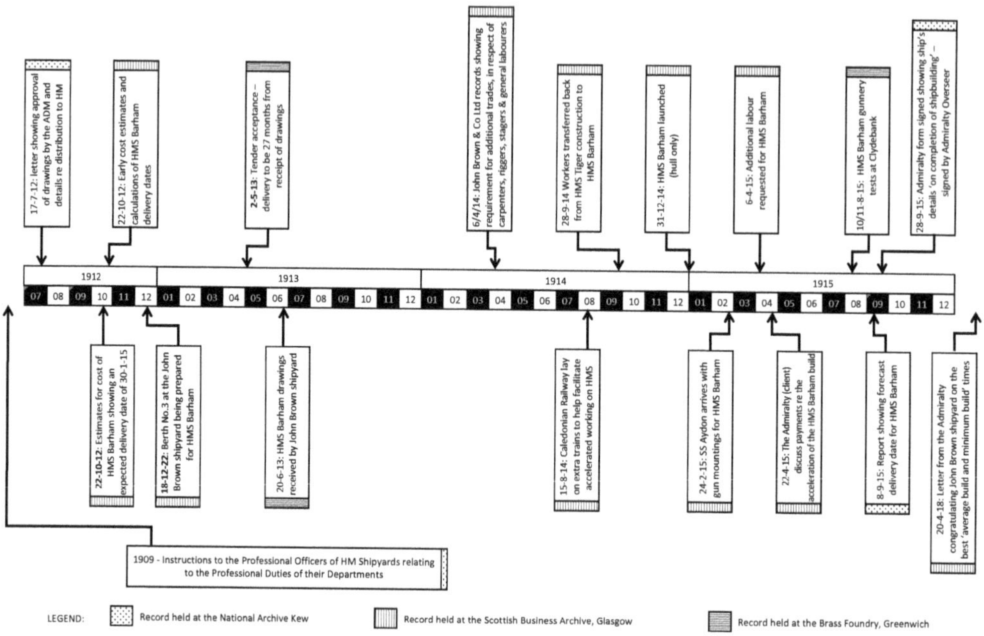

Figure 4.11 HMS Barham's project 'lifecycle' 1912–1915.

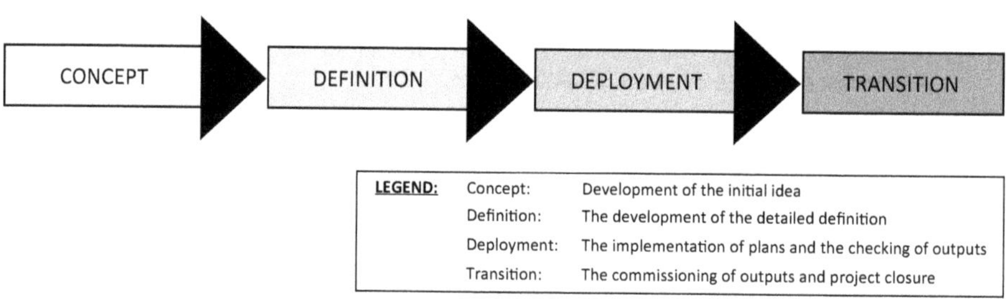

Figure 4.12 APM lifecycle BoK7.

References

1 APM. (2019). *APM body of knowledge* (7th ed.). Association for Project Management.
2 Project Management Institute. (2021) The standard for project management and a guide to the project management body of knowledge (7th ed.). Project Management Institute, Inc.
3 Ohara, S. (2017). *P2M – A guidebook of project & program management for enterprise innovation* (3rd ed.). Project Management Association of Japan (PMAJ).
4 Chartered Institute of Building Issuing Body. (2022). *Code of practice for project management for the built environment* (6th ed.). Wiley Blackwell.
5 Great Britain. Cabinet Office, & Sowden, R. (2011). *Managing successful programmes* (4th ed.). [authoring team, Rod Sowden et al.]. TSO.
6 Goetz, A.R., & Szyliowicz, J.S. (1997). Revisiting transportation planning and decision making theory: The case of Denver International Airport. *Transportation Research Part A: Policy and Practice*, *31*(4), 263–280.

7 Evans, G., & Veazey, R. (1998, February). Denver International Airport—The world's biggest. *Proceedings of the Institution of Civil Engineers-Civil Engineering, 126*(1), 2–11.
8 Johnson, K. (2005). Denver airport saw the future. It didn't work. *New York Times,* A1(L). https://www.nytimes.com/2005/08/27/us/denver-airport-saw-the-future-it-didnt-work.html [Accessed 10 July 2024]
9 Montealegre, R. (1996). What can we learn from the implementation of the automated baggage-handling system at the Denver international airport? *AMCIS 1996 Proceedings*. https://aisel.aisnet.org/amcis1996/327
10 Keil, M., & Montealegre, R. (2000). Cutting your losses: Extricating your organization when a big project goes awry. *MIT Sloan Management Review*, *41*(3), 55.
11 Airlines for America. (2024). U.S. airline traffic and capacity. https://www.airlines.org/dataset/annual-results-u-s-airlines-2/ [Accessed 10 July 2024].
12 Peebles, H.B. (1986). *Warship building on the Clyde, naval orders and the prosperity of the Clyde shipbuilding industry, 1889–1939*. John Donald Publishers Ltd.

Chapter 5

Project Roles and Organisation

A project is dependent not just on the quality of the people working on the project but also on how they interact together, as a team, and how they interact with their organisation as a whole. This chapter explores the purpose of project roles and the impact of different project organisations.

5.1 The Purpose of Project Roles and Organisation

Projects can be stand-alone enterprises, or they can be part of a larger programme or group of related projects within a larger corporate enterprise. P2M[1] says that where a project organisation is inserted into the framework of a larger enterprise then attention needs to be paid to the way in which the project relates to its parent organisation in order to ensure maximum project efficiency and effectiveness in the way in which it operates and delivers its outputs.

If a project is part of a large organisation, then it is more than likely that other projects will be being undertaken at the same time. The APM[2] say that multiple projects can run in parallel but where related or concurrent, they will require prioritisation in terms of scheduled delivery and the use of key shared resources including staff and finances. The project structure and organisation will need to reflect this.

Within a project team, there will be a range of different roles. The APM[2] and PMI[3] talk about roles such as project manager, project sponsor, the project team, users, governance board and product owner amongst others.

The roles above are all stakeholders (see Chapter 6). The term stakeholder is wide and within a project, is defined as all those, individuals or groups, who have an interest in the project or who are impacted by it. It is also worth noting that many of these roles can be broken down further, for instance a project team member could specialise in and be assigned to one of many tasks such as analysis, coding, testing or inspection of IT code being written. P2M[1] details similar roles adding that some, such as contractors will be outside your organisation, and those inside your organisation will have varying degrees of responsibility and authority. An example here is that within the project team itself, some will have direct responsibility for individual sub-sections of the WBS (see Chapter 8). These sub-sections or 'arms' of the WBS are often referred to as workstreams giving rise to the term 'workstream leader'.

How these roles function and are managed is linked closely to the organisation or structure of the project. Whether the project is a stand-alone enterprise or part of a larger organisation will impact the approach taken and the balance of power or authority which the project manager has.

DOI: 10.1201/9781003470748-6

5.2 How Do Project Roles and Organisational Structure Aid Project Delivery

5.2.1 *Organisational Structures*

Each of the roles discussed in Section 5.1 are defined in Table 5.1 below.

Table 5.1 defines the key roles within a project, there is a sense of hierarchy amongst roles 1–4 which will be analysed further in Section 5.2.2. There is also a sense of the two sides of the project, reflecting the costs and benefits discussed in the business case of Chapter 2, the costs will be incurred by roles 1–4 and these are justified by the benefits reaped by roles 5 and 6.

There are three main types of organisational structures within which projects can be delivered. These lie on a spectrum as shown in Figure 5.1.

Figure 5.1 shows the three main types of organisation, numbered 1–3, Functional, Matrix and Project. It also shows that within the category of the Matrix Organisation, there are three distinct sub-forms of organisation, a Weak, a Balanced and Strong Matrix.

Each one of these categories has a different structural organisation and each brings different strengths and focuses to the delivery of the project.

If a typical company layout was considered and drawn, then it is quite likely that a diagram such as that at Figure 5.2 would be revealed:

Here the company is organised by function, in this case, Sales, Operations, Finance and IT. Each area reports into the main company board, and each area has its own managers and members of staff focused primarily the aims of their departments. The structure represents the most common way in which companies have developed over time and one which provides a 'home' for functional experts. There is some advantage in this as difficult issues can be worked on together and it provides the opportunity for mentoring and developing the next generation of functional experts.

Table 5.1 Definitions of Project Team Roles

1	**Governance Board**	A body that provides overall direction sponsorship to a project. The board will represent financial, provider and user interests. Members of a governance board oversee project deployment and ensure that the project is governed, and delivered, effectively and effectively
2	**Project Sponsor**	A person (or group) who provide resources and support for the project and is accountable for enabling success
3	**Project Manager**	The person assigned by the performing organisation to lead the project team and take day to day responsibility for achieving the project objectives. This involves a number of functions, such as managing the project processes, delivering project outcomes, and enabling the project team to deliver project outcomes
4	**Project Team**	A group of people working together and collaboratively delivering project goals
5	**Product Owner**	A person responsible for maximising the value of the product and who is accountable for the end product
6	**Users:**	The group of people who are intended to work with, or use, the deliverables to enable project benefits to be realised

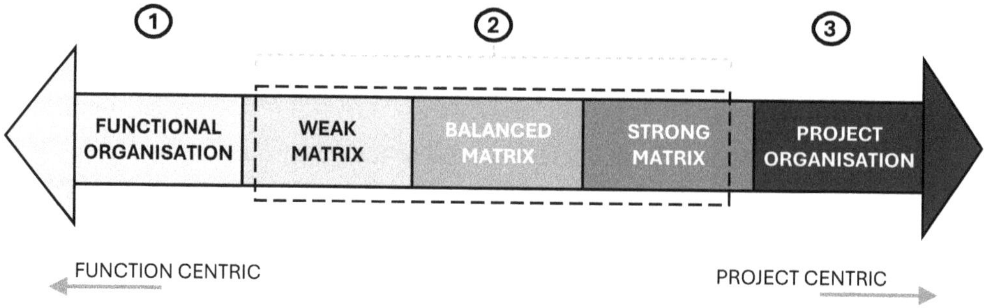

Figure 5.1 The spectrum of organisational structures.

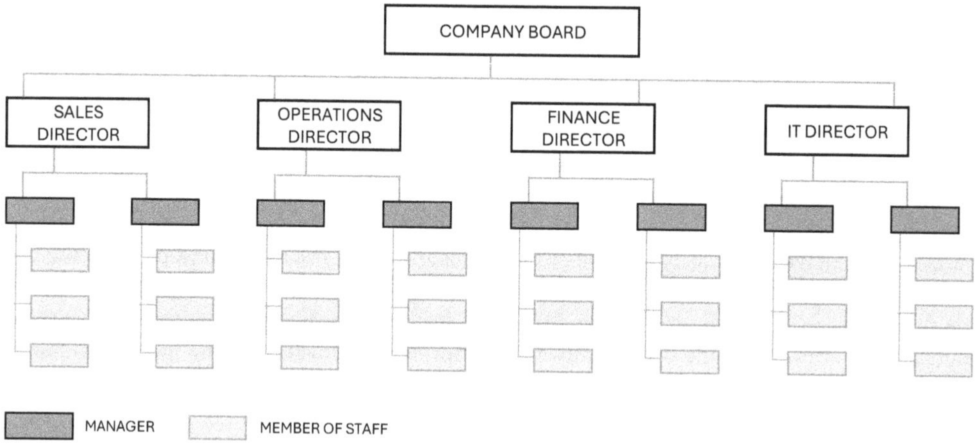

Figure 5.2 An example of a functional organisational structure.

The keen focus of each department can mean that sharing work between the departments is difficult. Staff members' hopes of promotion together with the day to day management of these staff, lie directly within each department or silo, meaning that working for a different department can become counter intuitive. This is not a problem for the organisation if all activity including projects, lies within a single silo. It would, however, be very rare to have an internal project without, for example, some input from IT, or without some impact on financial or operational issues.

All the departments within an organisation should be serving the bottom line of the commercial organisation and therefore share the interest of the other functions but this is not always the case.

There are however alternative structures that should be considered.

The most extreme of these, the polar opposite of a functional organisation, is shown in Figure 5.1 - the Project Organisation.

Figure 5.3 shows a project organisational structure where all work undertaken is delivered as a project. An example of this sort of scenario could be a consultancy firm where each new piece of work won is treated as a project where the costs and income associated with that project are collated and kept separate. The admin core provides internal services needed by all projects, such as HR, and which if provided by the individual projects would be uneconomic.

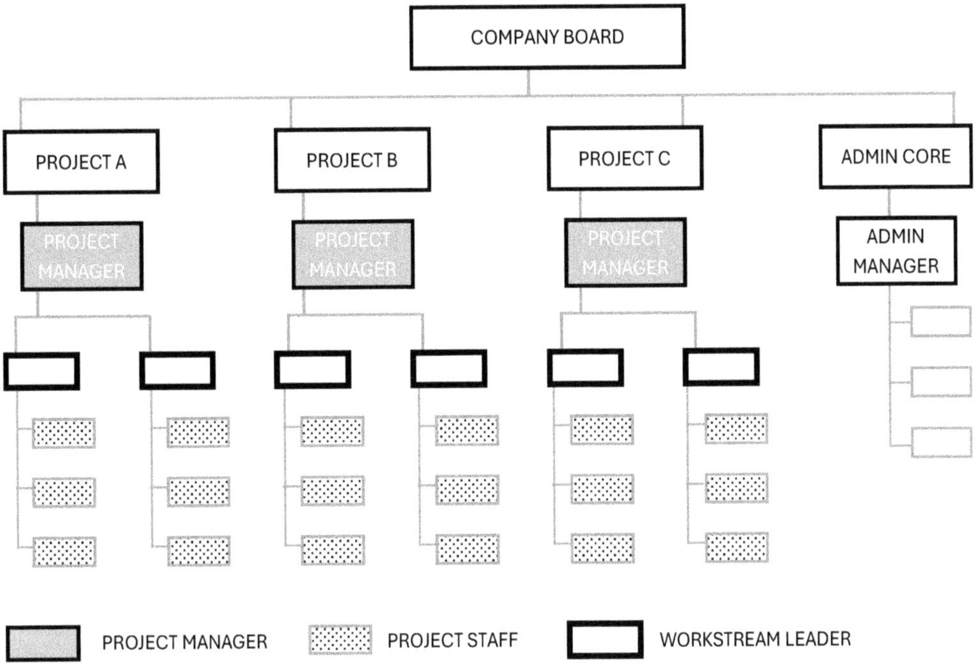

Figure 5.3 Project team members within a project organisational structure.

Each project shown in Figure 5.3 is managed by a project manager reporting into a project or programme director who sits within the main board of the company.

Exactly which structure is best suited to the organisation will depend on a number of considerations. Projects themselves can be quite simple and straight forward, in which case they could be delivered without recourse to a project specific structure. They may have few tasks and a limited number of specialist resources. On the other hand, projects may consist of multiple individual tasks involving a large number of different trades and specialists, thereby making it very difficult to deliver them within a standard functional organisation. An important question that needs to be asked here when deciding on the preferred structure, is whether the project is a sub-set of the organisation's activities, or whether it is a set of independent activities disconnected from the organisation's general functions that is to be delivered by a temporary team. For the organisation to be set up as in Figure 5.3, there would need to be a continual flow of projects, indeed projects would need to be the company's business. Organisations within firms inevitably establish physical and operational boundaries within the firm and so the organisational form needs to be well chosen in order to maximise the company's synergy and ability to function day to day. The form chosen will also help to define the relationships not just between team members of a project, but between projects and the environment outside of the projects.

The decision as to which of the organisational structures shown so far to use, is stark and offers a harsh choice of one end or the other of the spectrum shown in Figure 5.1. If this is the selection which has to be made, then it would be useful to review the advantages and disadvantages of each, as set out in Table 5.2.

Table 5.2 shows that it is quite possible that neither end of the spectrum is totally suitable for a company's aims. Very few organisations run without projects yet relatively few depend solely on projects to deliver their output.

Table 5.2 Advantages and Disadvantages of Functional and Project Organisations

	Advantages	*Disadvantages*
Functional Structure	• Stable centres of technical expertise	• No focus on project goals
	• Ability to learn from specialists	• Low authority given to lead projects
	• Clear lines of authority	• No sense of identity
	• Simple, straight forward structure	• Passed between departments
	• Clear career development path	• Any project work is an 'inconvenience'
	• Team focused on project goals	• Poor integration
Project Structure	• Project Manager has authority & responsibility	• Can be low technical expertise in teams
	• Ability to respond quickly to issues	• Often one expert assigned, so communication between multiple experts is limited
	• Good communications within the team	• Possible duplication of work
	• Less resource conflicts	• Difficult to stop projects soon enough if required
	• Potential career option	• Loss of team members to another team

A compromise, or mid-path, that takes the best of both extremes can be the best option in order to allow companies to maximise returns from their operational structure. These mixed organisations are called Matrix Organisations and can be structured to be Weak Matrices (in terms of the degree to which they are influenced by the project organisation,) Balanced Matrices or Strong Matrices which are heavily influenced by the project organisation.

The first of these matrices, the weak matrix is most suited when the organisation is predominantly focused on functional output as opposed to the delivery of projects. Projects will still be part of the day to day activity but will be of less frequent or of less importance than the rest of the organisation's business.

Figure 5.4 illustrates a weak matrix organisation. In this scenario the project is worked on by a number of team members located across multiple functions within the organisation and the unstructured nature of the diagram hints at some of the challenges for project delivery. The project is delivered by a Weak coalition of workers across a number of different functions. There is no single point of accountability, and the team is left to manage both itself and the delivery of the project. This could work for a small project or a project which has no sense of urgency. Once the project gains a higher profile and a greater level of importance within the organisation, it will be necessary to have a single point of accountability and responsibility in order to drive the project forward and to provide the hands-on co-ordination which a larger project will require. This leads to the balanced matrix structure shown in Figure 5.5.

In a balanced matrix, the project manager can be drawn from any level within the organisation but will still need to negotiate with each project team members' line management for the necessary number of staff hours required by the project. The staff member's time is rarely given 100% and so there is the potential for conflict, lack of resources and also the strong possibility that the team member may be more focused on their 'day job', this is after all where their promotional prospects lie. The project manager will also often need to take whichever staff are offered. The project, unless it has a high profile, is therefore unlikely to be given the best staff by the line manager who will want to keep them to deliver the functional department's targets.

The balance of power between functional and project deliverables is tilted towards the project environment in a strong matrix. Here the project has representation at board level and project

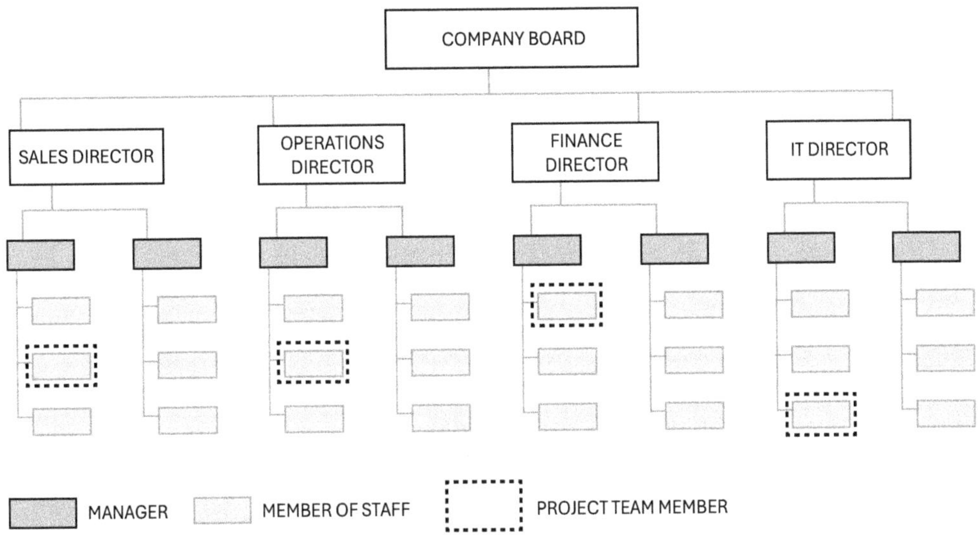

Figure 5.4 A weak matrix organisational structure.

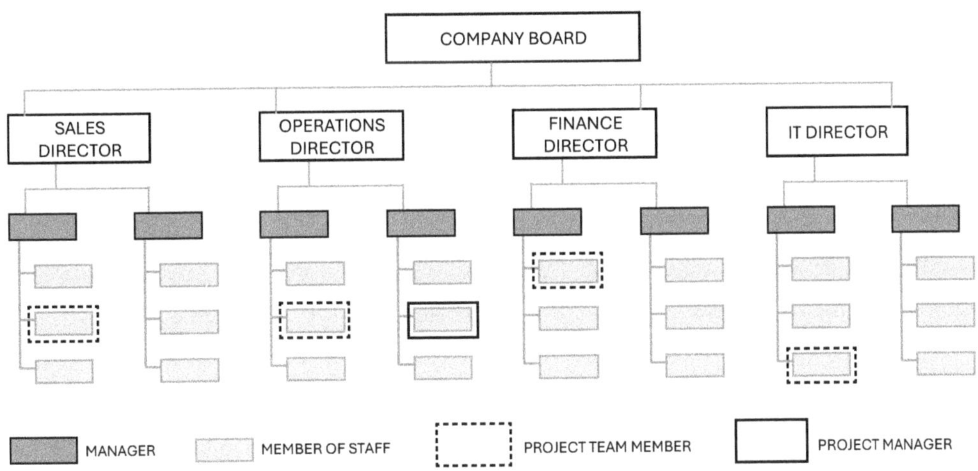

Figure 5.5 A balanced matrix organisational structure.

managers are taken out of the functional structure and placed within a specific project management department. Figure 5.6 illustrates how this type of matrix is set up. The project managers are removed from the functional structure and report directly into a Project Director who can represent project related issues at board level. The project managers can focus on what is best for the organisation as a whole rather than being subject to direction from one particular function as in the balanced matrix (see Figure 5.5).

Each of the organisational structures discussed have possible advantages for the organisations and each could be the best choice given certain circumstances.

Table 5.2 set out some advantages and disadvantages of both the functional and project organisations, Table 5.3 below addresses some criteria which might be used to help guide the choice between a functional, matrix or project organisational structure.

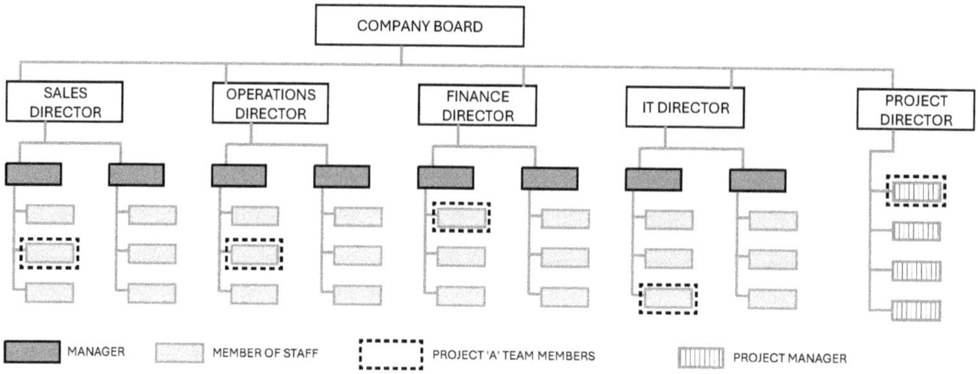

Figure 5.6 A strong matrix organisational structure.

Table 5.3 Project Criteria to Consider When Selecting an Organisational Structure

Project Criteria	*Functional Organisation*	*Matrix Organisation*	*Project Organisation*
Uncertainty	Low	High	High
Technology	Standard	Complicated	New
Complexity	Low	Medium	High
Duration	Short	Medium	Long
Size	Small	Medium	Large
Importance	Low	Medium	High
Customer	Diverse	Medium	One
Interdependency	Low	Medium	High
Time Criticality	Low	Medium	High
Differentiation	Low	Medium	High

5.2.2 *Project Roles*

Figures 5.2–5.6 have shown a number of structures and within them a number of project roles have been identified, but not defined. This is addressed in this section.

The first, most obvious role is the project manager. This role is responsible for the planning of the project, the development of a delivery schedule and the management of the project team's workload through the project's life. They are responsible too for the successful delivery of all project processes, as discussed elsewhere in this book, such as risk and budget management etc. Working with the team they will drive the project forward providing the overall day to day direction and reporting on progress both to the team and upwards towards the project sponsor.

The project team members are those that deliver the tasks defined in the Work Breakdown Structure (see Chapter 8), and as assigned by the project manager. The team members may include a workstream manager who will have specific responsibility for one area, or arm, of the Work Breakdown Structure. The team members are accountable to the project manager, or workstream manager if in place, and will update them on progress and task completion.

The project sponsor owns the business case (see Chapter 2). They are accountable for the benefits which flow mainly in the in-service part of the project's extended lifecycle. These need to be harvested in order to be able to offset them against the project's costs or finances, for which the project sponsor also holds ultimate responsibility. The project sponsor will, in an internal project, usually be someone in a higher position within the organisation's hierarchy, and consequently will be in direct contact with the project's main stakeholders and the project manager will frequently report directly to them as well as their own line manager.

A steering committee, consisting of a cross-functional group of senior management, may also exist within the project hierarchy. Their purpose is to act as an advisory panel helping to steer the project, ensuring that the project is aligned to strategic goals and to provide a direct link to the business.

These four key roles are responsible for the escalation and delegation of decisions and direction for the project, as illustrated by Figure 5.7.

Figure 5.7 shows how the main roles within the project will need at times to refer issues to the next level in terms of seniority. When the team members have an issue or problem that they can solve then they do so. Where they can't do so, either due to complexity or the degree of impact on the project's timescale, costs or quality of output, then this needs to be referred up to the project manager. The project manager will have a level of authority which allows them to make decisions on issues with an impact up to a certain threshold. If this threshold is not reached, then the project manager will issue revised tasks as required and pass these back to the team. Where the threshold is breached then the issue, with a defined solution wherever possible, should be referred to the project sponsor. If needed, the project sponsor can take the issue to the project steering committee, but either way revised requirements as necessary will be passed back to the project manager for dissemination amongst the team.

There are of course other project roles such as Project Administrator (providing support with administrative tasks, such as reporting and placing orders for project components), Project Analyst (responsible for gathering and analysing data for the project), Change Control Board (see

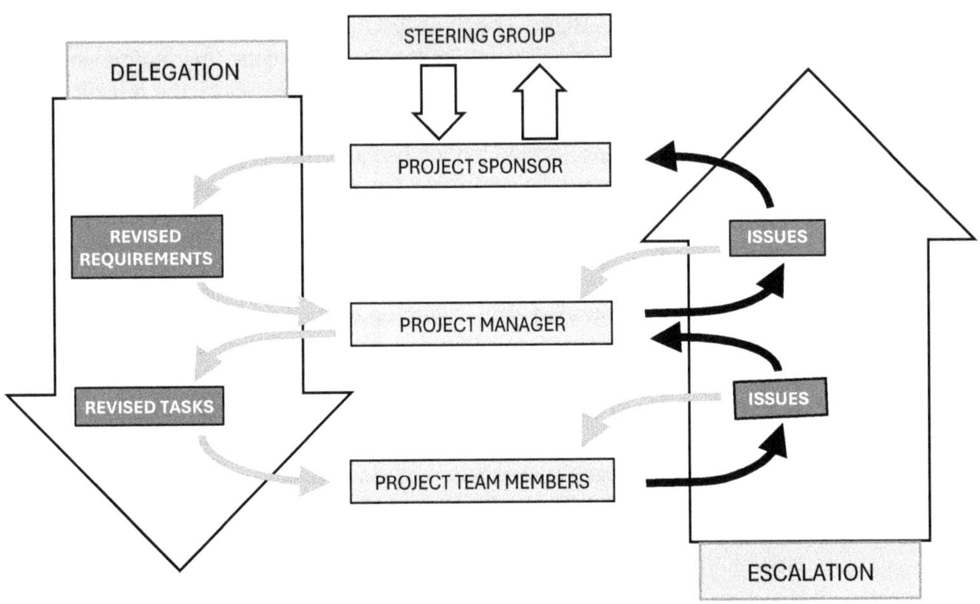

Figure 5.7 Project decision making – The escalation and delegation process.

Chapter 12) and a team of end users consisting of those that will use the project outputs to gain the project benefits. A Benefits Manager may help to map the benefits (see Chapter 12,) and ensure that the end users are able to maximise the gains from the project. A Project Management Office can also be key, and they will provide policies and standards applicable to the organisations project delivery and maintain project records. They can also act as a depository for project reports from previous projects and can hold records that could be of use in, for example, the estimating process (see Chapter 7).

These later roles are more likely to be present the larger the project or organisation is.

5.3 Project Review: The impact of Project Roles and Structure on the Project: Syndey Opera House

5.3.1 *The Project Background*

Sydney Opera House is known throughout the world as a landmark on the waterfront of Sydney Harbour. Yet the project was troubled and needed to overcome numerous difficulties in order to be successfully completed. In this section the impact of the project roles and organisation are discussed.

The Australian Broadcasting Corporation (ABC) played a key part in the history of the Opera House. In the 1950s it was Sir Eugene Goossens, who was the conductor of the Sydney Symphony Orchestra, who lobbied hardest for an Opera House to be built in Australia, but it was Sir Charles Moses, then head of the ABC, who introduced Goossens to Premier Joseph Cahill, a key meeting that helped crystallised the idea.[4]

A conference in November 1954, focused on the establishment of an Opera House in Sydney where the then Premier Joe Cahill announced the formation of a committee to examine sites, plans and designs,[5] showing that the Premier was a key figure in developing the project from the project's early days. The committee ratified the winning design in 1960.[5]

From an initial forecast of A$7m in 1957, the project costs ballooned by a factor of 14.5 to an actual outturn cost in 1973 of A$102m. The Sydney Opera House Act of 1960[6] had stated that the 'estimated cost shall not, under any circumstances, be exceeded by more than ten per centum', so it is clear to say that the rate of cost increase encountered by the project was far from expected!

The project schedule was also impacted and having had an original completion date of January 1963, it took a further ten years until the completed Opera House was able to open in 1973.[7]

5.3.2 *The Project Issue*

The project was slow to get off the ground. The Danish architect Jørn Utzon,[8] 1 of 233 entries, won the design competition conducted by the NSW government in 1957 for the new Opera House with an unusual design based on the sails of a ship, and gull wings which drew on elements of ancient Chinese architecture.[9] The sponsor of the project was the Labour Party Premier of NSW, Joe Cahill, who in order to gain approval for the project is said to have understated the construction costs.[9] The project therefore began with a very unusual, almost revolutionary, roof design and an insufficient construction budget. The construction was begun by Arup in 1959, with a tight four-year project schedule. It took Utzon and Arup until 1961 to work out how to construct the now iconic roof. Then shortly after, Utzon resigned from the project and was replaced by a board of four new architects. As a result, a key team member – the winning architect Utzon, was no longer working with the team and the vision for the project had to be re-established.

The strong support from Cahill was given despite political opposition, indeed he was noted as having 'done more than anyone to push the project, especially after 1956 when Sir Eugene Goossens left Australia. As a result Cahill spent a great deal of political capital in the process'[10] of pushing the build forward. The construction of the Opera House became Cahill's pet project, he effectively took the project under his arm and ran with it.[11] As Premier of Australia, it is questionable whether he had sufficient time, or construction knowledge to devote to the project. He did so without the use of an end user team or owners' team who could have helped to steer the project. An example of how this could have been beneficial is that early consultation with them could have avoided some of the late design changes that ended up adding cost and time to the project. One such change meant that the major hall originally built to be a multipurpose concert and opera hall was redesignated as a purely a concert hall. This meant that A$5m worth of stage machinery already purchased had to be written off.[11]

Having taken such as key role, without the support of standard project roles that we would use today, disaster then struck on 22 October 1959 when Cahill died at the age of 68. Before he died, Cahill asked Norman Ryan, his Minister for Public Works, to promise not to let the Opera House project fail. Ryan, who had claimed that he didn't understand the Opera House, struggled to properly manage the enormously complex building project,[12] eventually delivering it in a revised form 14 years after the first stone had been laid by Cahill in 1959.

5.3.3 How Could More Developed Project Roles & Organisation Have Helped?

If the project had employed a professional project manager, then there would have been a far greater focus on risk management, Cahill himself did not undertake any risk management.[11]

The lack of clarity over the design meant that the road to the final result was far from straight forward, meaning that the project spent both time and money where it was not required.

There is the suggestion that Cahill attempted to fulfil the roles of both project manager and project sponsor, these roles need to be separate in order to allow for constructive challenges to be made and for the correct checks and balances to be in place. 'Pet Projects' are always particularly dangerous and have a tendency to overspend and/or not be stopped soon enough when going wrong, something which inevitably adds to the amount of resource consumed by them.

The benefits of the project would have been clearer and more fully understood, had a Benefit's Manager been appointed to focus on these issues (see Chapter 12). The project's seeming lack of consultation and engagement with end users is a major omission and contributory factor to the overruns. When necessary changes were eventually identified, much later than they would have been had end users been fully involved, then the costs of implementing the changes were increased as previous, nugatory work had to be undone and rectified.

The project allowed others who were not experts to 'meddle' in the design and construction of the Opera House with unsurprising impacts on the project costs and risk profile. Key parties such as the civil engineers and architects were changed without consideration of the impact[11] on the project's delivery continuity. A formalised project structure would have enabled a better focus on these issues with the impact of each being fully understood and mitigated before being allowed to proceed.

The centralisation of project information was also an issue for the project. Tellingly, when Cahill passed, he gave responsibility to Ryan the Minister for Public Works. In Cahill's mind this was the best person for the job, yet Ryan's own admission was that he didn't understand the project. A more considered project organisation would have made this transition, and that between Utzon and the replacement architects, less troublesome.

5.3.4 *Conclusions on the Sydney Opera House Project*

Sydney Opera House from a project management perspective was a disaster. Progress on the project became 'a joke' and the reporting of this added additional pressure to Cahill.[11]

Despite the lack of a project structure with individuals fulfilling project roles in a way that meant the project was weakly managed, despite there being no risk analysis, prioritisation or even understanding of the risks, and despite there being little control of design or budgetary control, the project is now widely viewed as a success. In monetary terms it is a success, if all the now acquired benefits are counted. The costs rose astronomically to become far greater than ever intended, but so too did the level of benefits. The benefits to Australia of building an iconic building such as the Opera House are similar to those that London gains through the Tower of London, or those that Paris does through the Eiffel Tower.

If project costs, although substantially understated in the business case and poorly managed throughout the lifecycle of the project, are compared to the full benefits generated, a picture of project success does emerge. The benefit to Australia over the years in terms of tourist dollars due to the international profile given by the Opera House, although hard to substantiate, must far outweigh the costs involved in construction. The benefit however could have been greater had the project management been more focused and if some of the approaches discussed above had been used. The success of the project is entirely dependant on the benefits, largely linked to profile and tourism, which were not considered when Sir Eugene Goossens, Sir Charles Moses, and Premier Joseph Cahill first contemplated the project. The project's success could therefore be viewed as a result of good fortune rather than good project management!

5.4 Project Roles and Organisation Before the Birth of Modern Project Management*

(* In this section, the research relating to the construction of HMS Barham 1912–1915 mentioned in the introduction is drawn upon to help illustrate how the subject of this chapter has developed.)

In this section, the text returns to the consideration of project delivery prior to the beginning of what is known as Modern Project Management. In the early part of the twentieth century, the British Navy had a programme of shipbuilding each year. The programmes would consist of approximately four or five major vessels, and each was run along lines similar to those of a modern project. In terms of project organisation, the first ship in a programme was more often than not built in one of HM Dockyards but after that the ships would be tendered out to private shipyards. HMS Barham was the fourth ship in the 1912–1913 Programme consisting of HMS Queen Elizabeth, HMS Warspite, HMS Valiant, HMS Barham and HMS Malaya. As such, HMS Barham was constructed at the John Brown shipyard on the Clyde. The dockyard workers building HMS Queen Elizabeth in HM Dockyard Portsmouth were governed by detailed definitions and instructions as to what was expected from each role. These were defined, for HM Dockyards, in a book titled '1909 Instructions to the Professional Officers of His Majesty's dockyard at Home in Matters Relating to the professional Duties of Their Departments Together with Directions Relative to Building, Fitting, and Completing Ships, Defects, Alterations and Additional Fittings in Ships and Estimates and Expense of Work etc.' The lengthy title page is shown below in Photo 5.1.

The sheer length of the title, at Photo 5.1, is an indication as to the depth of detail contained in the guide.

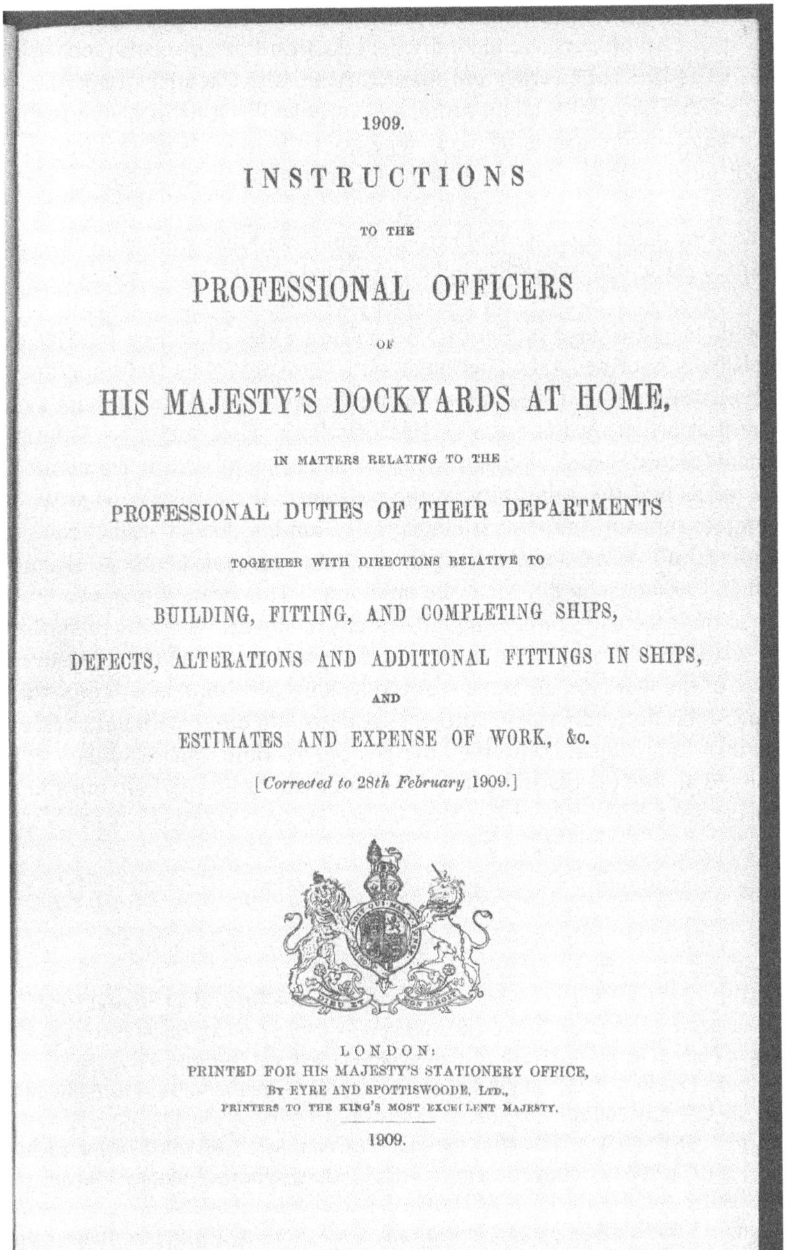

1909.

INSTRUCTIONS

TO THE

PROFESSIONAL OFFICERS

OF

HIS MAJESTY'S DOCKYARDS AT HOME,

IN MATTERS RELATING TO THE

PROFESSIONAL DUTIES OF THEIR DEPARTMENTS

TOGETHER WITH DIRECTIONS RELATIVE TO

BUILDING, FITTING, AND COMPLETING SHIPS,

DEFECTS, ALTERATIONS AND ADDITIONAL FITTINGS IN SHIPS,

AND

ESTIMATES AND EXPENSE OF WORK, &c.

[*Corrected to 28th February* 1909.]

LONDON:
PRINTED FOR HIS MAJESTY'S STATIONERY OFFICE,
BY EYRE AND SPOTTISWOODE, LTD.,
PRINTERS TO THE KING'S MOST EXCELLENT MAJESTY.

1909.

Photo 5.1 Title page of the 1909 HMSO publication governing roles within HM Dockyards. By permission of The National Archives of the UK: ADM 95/107.

The book extends to 20 chapters and its 320 pages cover subjects ranging from the detailed duties of key dockyard officers and the individual dockyard departments (such as Engineering, Construction, Electrical Engineering and even the Admiralty Chemist's Department).

Further chapters address 'Building, fitting and completing ships for commission', 'Progress of work', 'Estimates and expense of work' together with appendices detailing the 'Periodical accounts reports and returns' which each of the four departments listed above, together with the Captain of the Dockyard, were expected to file.

Not only does the publication set out the roles within HM Shipyards, but it also establishes the manner in which particular tasks were to be completed and the records etc that were expected to be kept.

Even though the title page at Photo 5.1 is dated 1909, the copy held at the National Archives in Kew is inscribed, on a later page, with the words 'W J Dally foreman of the yard 12/4/15' demonstrating that it was in use during the period of HMS Barham's construction. The ships built at private yards, such as HMS Barham, were subject to Admiralty inspections and many records exist of discussion of changes required or payments to be made between the yards and the Admiralty as the customer, the client or what we would term today, the project sponsor. In the next chapter, Section 6.4 details a luncheon that was held in the Moulding Loft of the shipyard for foremen and several members of the staff on the launch of HMS Barham with a ', Mr Luke presiding'. This bears witness to the structure of the 'team' that built the ship, indicating a hierarchy as not all members of staff who worked on the ship were invited to the luncheon. The foreman mentioned would have held a role similar to that of the workstream leaders of today, and 'Mr Luke' as a senior role within the yard took on similar duties to today's project manager. No organisational chart exists, either it was not drawn up or has not survived the passage of time, but secondary evidence such as that discussed above, clearly indicates that comparable roles and organisation to those of today's projects did exist.

References

1 Ohara, S. (2017). *P2M – A guidebook of project & program management for enterprise innovation* (3rd ed.). Project Management Association of Japan (PMAJ).
2 APM. (2019). *APM body of knowledge* (7th ed.). Association for Project Management.
3 Project Management Institute. (2021). *The standard for project management and a guide to the project management body of knowledge* (7th ed.). Project Management Institute, Inc.
4 Sydney Opera House. (2023). http://theoperahouseproject.com/ie/transcripts/A-Serendipitous-Idea.htm [Accessed 25 July 2025].
5 Government of NSW. (2015). GY-6401 Opera House Committee (1954–1957) Opera House Executive Committee (1957–1961). https://search.records.nsw.gov.au/primo-explore/fulldisplay?context=L&vid=61SRA&docid=ORGANISATIONS1005863 [Accessed 10 July 2024].
6 Government of NSW. (2010). Sydney Opera House Act. Act No. 2 9, 1 9 6 0. https://legislation.nsw.gov.au/view/html/1960-04-19/act-1960-029 [Accessed 8 July 2024].
7 Hale, P., & Macdonald, S., 2005. The Sydney Opera House: An evolving icon. *Journal of Architectural Conservation, 11*(2), 7–22.
8 Sydney Opera House. (2019). For demonstrating architecture's power to refine a city (most influential projects: #50). *PM Network, 33*, 94–95.
9 Bourne, L. (2007). Avoiding the successful failure. Paper presented at PMI® Global Congress 2007—Asia Pacific, Hong Kong, People's Republic of China. Project Management Institute.

10 Sydney Opera House. (2024). Our story – Construction begins. https://www.sydneyoperahouse.com/our-story/construction-begins [Accessed 10 July 2024].
11 De Beer, C. (2016). Risk management in the construction sector – Part 1: Case study. https://www.youtube.com/watch?v=kkM9BUTQMWc [Accessed 4 September 2024].
12 Sydney Opera House. (2024). Chapter 8: 1959, laying the first stone and the death of Cahill. http://theoperahouseproject.com/ie/transcripts/1959-Laying-The-First-Stone-And-The-Death-Of-Cahill.htm [Accessed 25 July 2025].

Chapter 6

Project Leadership and Soft Skills

Within project management, there are a number of what are termed 'hard skills' and a number of aspects termed 'soft skills'.

As might be expected, hard skills are comparatively tangible, they lead to definite project outputs such as a business case, a budget, or a risk register. Soft skills are less tangible and consist of competencies such as project leadership, teamwork, and negotiation. Conquering soft skills alongside the hard skills is however what can make the differences between a good project manager and an excellent one. EI or emotional intelligence fits closely with soft skills, whereas the hard skills are more aligned to standard measures of IQ.

Soft skills are harder to explore as they are often best discussed when related directly to project examples. Section 6.3 will perhaps help in this, but in Section 6.2 the general theoretical options are discussed.

6.1 The Meaning of Leadership, Communication, Negotiation, Conflict Management, Stakeholder Management and Teamwork Within a Project

The APM[1] define leadership as requiring the delivery of vision and direction to influence others and to align them with project direction, thereby delivering the project efficiently and effectively. The PMI[2] make the observation that leadership is not limited just to the project manager but should be displayed throughout the project team and should include elements alongside project leadership, such as technical leadership and team leadership. The PMI describe leadership as one of the 'general management skills' along with communication and negotiation which are essential to effective project plan execution.' Communication and negotiation, along with items such as conflict management and stakeholder management are competences included in what would generally be considered the list of soft skills.

Communication is often thought of and described as the process of exchanging information but is also importantly acknowledged as a process of confirming that there is a shared understanding of that information.[1] The information must be provided in a timely fashion and the management of it requires a focus on communications planning, information distribution, performance reporting, and administrative closure.[2] P2M states that the purpose of communication within the project is to ensure 'smooth delivery' of the project,[3] which could be rephrased as efficient and effective project delivery. The need for smooth delivery of a project is at the heart of the P2M statement that underlines the importance of the need to 'promptly gather information on the project's actual status through (the) communication (process) and to communicate this status accurately to stakeholders to help solve the various issues that arise in the course of the

DOI: 10.1201/9781003470748-7

project.' P2M also highlights the need for communication outputs to enable predictive management to prevent problems before they actually occur.[3]

Negotiation within projects is characterised as 'conferring with others in order to come to terms or reach an agreement'[2] and it can be linked to any aspect of the project, not just procurement or scheduling for example. The essence of negotiation has been described as the art of putting forward your aims and through a strategy persuading the other party to accept your stance. The best strategy for achieving this is defined as one where both parties reach an agreement through a win-win approach.[3]

Conflict management or conflict resolution is defined by the APM as the 'process of identifying and addressing differences that if left unmanaged would impact the successful completion of objectives',[1] as such it can be seen to share some approaches used within negotiation. It is a key general Project Management skill, which used well can enhance team performance.[2] It has been said that it is important to avert problems before they cause a conflict,[3] but this chapter also addresses the potential benefits of controlled, or positive conflict within a project.

Stakeholder Management has four distinct areas, the identification, analysis, planning and implementation of actions designed to influence project stakeholders.[1] Project Management itself is described by the PMI as 'the application of knowledge, skills, tools, and techniques to project activities in order to meet or exceed stakeholder needs and expectations from a project'.[2] Engagement with the stakeholders is therefore paramount to project success and needs to be managed. Some ways in which various key stakeholders are engaged with have already been discussed in earlier chapters. Stakeholders include individuals impacted by the project or those who have an interest or role in it.[1] This would include the project sponsor who's needs will have been captured in the business case (see Chapter 2) and the project team some of who's key needs are covered later in a number of chapters, for example Chapters 8–10. The stakeholders within a project are many, meaning that different approaches are required to manage each group and it is this, in particular, that the management of stakeholders needs to focus on.

The last of the soft issues to be considered here is teamwork. The APM's BoK7[1] does not define 'teamwork' itself but establishes that a team needs to collaborate and work towards common goals. Part of the challenge for the project manager is that the team is temporary, reflecting the nature of all projects, and needs to reach optimal performance as quickly as possible if the project is to minimise costs and maximise benefits. The PMI[2] reference 'team development' as being essential to 'developing individual and group skills to enhance project performance' which provides an excellent description of the purpose of teamwork, as referred to in this chapter. P2M[3] adds some detail to this, stating that the main objectives of team building are to enable the team to function effectively by ensuring that the team has a common, and correct, understanding of the project objectives and how these have been developed leading to their expression in terms of time, cost, quality, risks and intended benefits.

6.2 How Are Soft Skills Delivered and Managed Within a Project?

This section provides an overview of each of the six soft skills outlined above from a project manager's perspective detailing some of the key aspects of each. There is of course much more to be said about these issues, but this section serves to introduce the approach to each topic and to give the project manager some initial areas to consider when running a project.

6.2.1 Leadership

Project managers are managers. They are also leaders. What does this mean and what is the difference between the two roles?

The PM is expected to furnish a number of management deliverables on behalf of the project. Some of these are the ability to plan and budget for the project, to organise the project staffing, to monitor and control project progress and to take an active and pre-emptive approach to risk management and problem solving. All these aspects help to produce a degree of order and predictability for the project and can help meet the short-term needs and demands of project stakeholders.

Project leadership on the other hand delivers change that stakeholders need in order to make the organisation fitter for the future and more competitive. This means that the project leader needs to employ a selection of softer skills to assist the project staff to deliver the project, such as those focused on enabling people, directing them, and influencing the team so that they are aligned with the organisations goals both in the short and longer term. In contrast to the leader, managers will tend to emphasise the 'hard' skills more, setting clear targets making short-term decisions and solving the short-term problems.

The project manager needs both these approaches, they need to have one eye on the day to day running of the project and one eye on the future. The management skills of the project manager are covered in a number of chapters, such as Chapters 4, 6, 8–10, the focus of this chapter however is solely on the approach to leadership itself and the basic choices, approaches and considerations available to the project manager.

If the characteristics of good leaders through history were to be reviewed, then doubtless qualities such as being an effective communicator, being a long-term thinker, self-motivated and confident whilst also being a 'people person' able to inspire and direct others would come up. Equally if a list of characteristics not required by leaders were to be fashioned, things such as a fearful approach, particularly of change, being dismissive of the ideas of others and unable to empathise, being indecisive and blaming others rather than accepting responsibility, would surely feature on that list. The CIOB[4] undertook a survey looking at leadership in the construction industry and the Figure 6.1 below shows the results based on two key questions from the survey.

Over the years there has been debate as to how a good leader is 'made'. Trait Theory would suggest that there are various skills or traits a good leader will have. Figure 6.2 below illustrates some of these.

WHAT ARE THE TOP 5 LEADERSHIP QUALITIES?

	QUALITY:	COMPARATIVE IMPORTANCE
1	Communication Skills	38%
2	Decision maker	19%
3	Integrity	16%
4	Ability to Listen	16%
5	Understanding	12%

WHAT ARE THE BOTTOM 5 LEADERSHIP QUALITIES?

	QUALITY:	COMPARATIVE IMPORTANCE
1	Not having enough time	38%
2	Don't delegate enough	19%
3	Impatience	16%
4	Poor communicator	15%
5	Lack of decisiveness	12%

Figure 6.1 Good and bad leadership qualities.

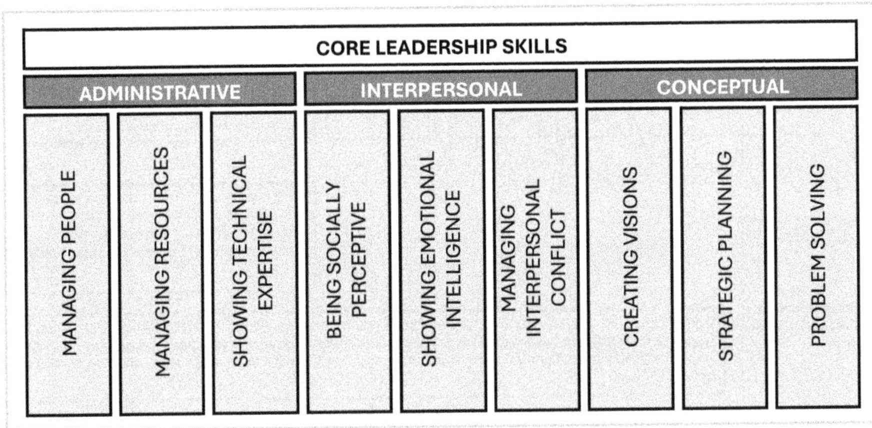

Figure 6.2 Examples of leadership traits or skills.

If a project manager were to be selected purely using trait theory, and whether they acquired the traits listed in Figure 6.2, either innately or through direct acquisition, then it is likely that other important aspects of their leadership approach would be overlooked.

Leadership Styles have also been deliberated upon for many years and three key styles or approaches, have emerged, as illustrated in Figure 6.3.

With multiple leadership styles available to the project manager, the question of which style is best suited to the project environment inevitably arises. The answer is, 'it depends'.

It depends on a variety of factors, such as the position of the project in its lifecycle, or how experienced the project team is or how well the project is delivering. The project will place quite different demands on the project manager if it is in difficulty as opposed to running smoothly. One style of leadership is not always going to deliver the best results for the project, which is encapsulated in the approach of 'situational leadership' where dependant on certain factors the ideal style of management can change in order to deliver the best results for the project.

An authoritarian leadership style can be the best choice where the project has considerable urgency, levels of uncertainty, is in a crises or is experiencing high levels of inertia. It is not generally favoured by the 'led' and can be disastrous if for example, the led don't share the same view of the 'crisis'.

Democratic leadership on the other hand is usually welcomed by the led and provides a wider decision-making base which can be good within a standard delivery phase. It can on occasion

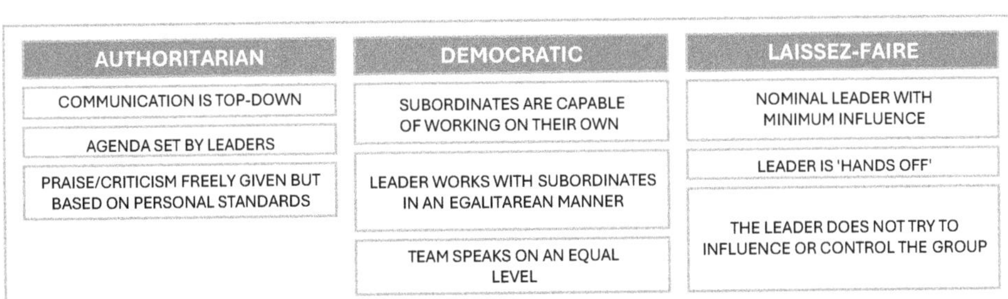

Figure 6.3 Examples and characteristics of leadership styles.

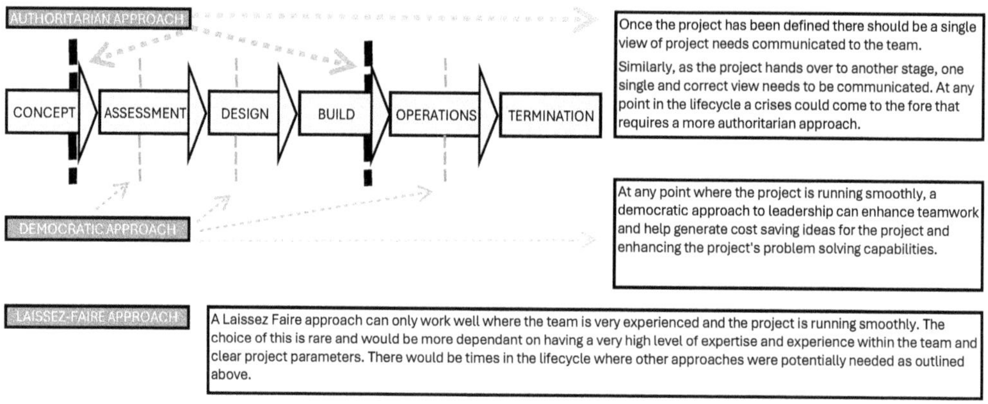

Figure 6.4 Suitability of leadership approaches through the project lifecycle.

bring to the fore a number of differing views meaning that an authoritarian style is needed in order to break the deadlock!

A Laissez-Faire approach seems perhaps one which should be avoided, yet with a very experienced and focused team it can be a valid option to let the team 'just get on with it', although it would be sensible to keep an eye on progress and ensure that it doesn't stall thereby using a light-handed management style approaching a Laissez-Faire approach. It can be seen then that the ideal leadership style will vary across the project lifecycle according to the situation the project finds itself in. Some of these considerations are set out if Figure 6.4 above.

6.2.2 *Communication*

Good communication is vital to any project. Without it there is the inevitable danger that the project team do not all pull in the same direction creating friction that will add to project costs and potentially reduce the level of benefits which the customer receives at the end of the project.

Successful communication within a project can be described as the ensuring of the exchange of relevant information. Within this description there are perhaps two words that need further explanation – relevant and exchange. If the information is to be relevant, then the audience for each communication needs to be considered, as does the aim or purpose of the communication and the way or method that is best used to achieve successful communication. In terms of progressing a project, relevant information will consist of items such as (although far from exhaustive):

- Project objectives
- Project deliverables
- Project plans
- Project benefits
- Project risks

Miscommunication of any of these will lead to increased project costs and potentially duration, so it is vital that communication is effective. One way in which this can be addressed is through a Communication Plan (see Figure 6.5) which has elements of the topic of stakeholder management which will be addressed later in this chapter.

A Communication Plan forces the project manager to think about the audience, impact, medium of communication and the potential side effects of the communication, giving the acronym AIMS.

MESSAGE	PURPOSE	AUDIENCE	COMMUNICATION OWNER	MEDIA	TIMING
PROJECT STATUS	PROVIDE INSIGHT INTO PROJECT PROGRESS AND ISSUES	CLIENT MANAGER	PROJECT MANAGER	WRITTEN REPORT	WEEKLY
PROOF OF FUNCTIONALITY	DEMONSTRATE THAT THE SYSTEM PROVIDES SAFE FUNCTIONALITY	OPERATOR AND SAFETY ENGINEER	LEAD PROGRAMMER	TEAM MEETINGS & SIGNED OFF FUNCTIONALITY RPT.	WEEKLY AND PRIOR TO ACCEPTANCE

Figure 6.5 An example of a project communication plan.

The information required for these stakeholder communications will need to be conveyed both precisely and correctly meaning that one, up to date, picture of the project and its progress is both given and maintained. The information contained within project communications can come from multiple sources, such as the Project Sterring Group (project brief/capital budgets/approved changes/project strategies etc.) the Project Team (agreed costs and progress against agreed plan/review of tenders received/design progress/changes required/progress against schedule/risk profiles etc.) the Project Design Team (implementation of matters relating to design and cost/general overseeing of design matters etc.) the Project Finance Team (the monitoring and reporting of cost related matters/preparing and advising of budgets/review tax matters/review costs implications of proposed changes etc.) As can be seen the sources of information can be many and varied just as the purpose and recipient of the communication can be.

In terms of how information can be communicated, there are three generic choices. The communication can be written, auditory or tactile. Each has its place within the project environment and the choice between them is often driven by the level of formality required, often linked to the need to keep a project record. These three different methods of communication and the format of that communication is illustrated in Figure 6.6. It can be seen that just because an initial

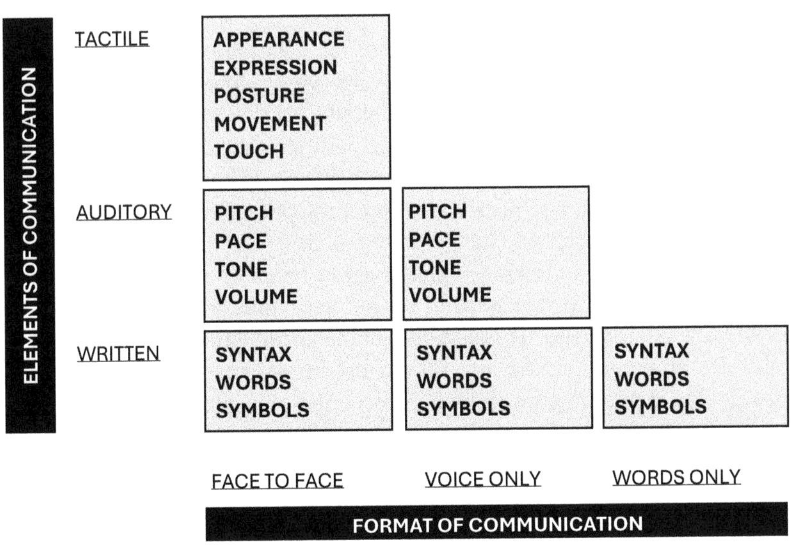

Figure 6.6 Elements and formats of communication.

tactile approach could be chosen, it does not preclude a formal record being created through a follow up email for example.

When a communication approach is selected, thought should be given to which method is used, particularly in terms of how the 'exchange' part of the earlier definition of communication – 'ensuring of the exchange of relevant information' is facilitated. In considering this, communication needs to be recognised as a two-way process and needs to include positive confirmation from those receiving the message. This can be delivered through feedback loops that confirm that the communication has been understood and interpreted correctly. Without confirmation that the communication has been received and understood there is a danger that the message may be mis-perceived, mis-interpreted or mis-evaluated. Any flaws in communication, such as these mentioned will lead to team members potentially seeing things differently, drawing different conclusions and deciding on differing actions. The more important, or central the communication is to the project, the more vital it is to ensure that the communication message is not just transmitted, but received, understood and recorded, if the project is to deliver efficiently.

6.2.3 Negotiation

The aim of negotiation within a project is to search for agreement or consensus and to ensure that views are aligned so that the project, or elements of the project, can go forward in one, correct, direction. This happens informally through the project lifecycle as well as in more formal settings such as discussions linked to procurement or to resource allocations.

Any negotiation needs to be planned for, and consideration should be given to the following four stages:

1 Preparation for the negotiation
2 Opening position set out
3 Bargaining within the negotiation
4 Closing the negotiation and recording agreement

The first of these stages means that in advance of the negotiation the project manager should consider what they want to achieve through the negotiation. In other words, they need to have clear objectives - what does the ideal end state look like? It is worth considering what the objectives might be for the other parties involved. This will include identifying which variables are likely to be discussed and any potential concessions that could be made by your side if required.

Consideration of the views and priorities of the project team are key inputs at this stage as is the potential power that your side has in the negotiation. Both of these aspects will help to determine the ideal approach or strategy that should be taken into the negotiation.

Once the negotiation begins it is wise to set out and agree the process and purpose of the negotiation with the other parties. This should include introductions from each side and a statement of both sides' objectives. An Agenda should also be agreed to give clarity around the issues to be discussed. The bargaining phase will involve the presentation of offers and evaluation and discussion of responses to those offers. There will usually be counteroffers and trade-offs accepted and rejected until an agreement can be reached. The final agreement should be summarised and details formally circulated to those present immediately after the meeting.

There are some key negotiating skills which the project manager will need to be aware of and to be comfortable using, a number of which are summarised in Table 6.1 below.

Table 6.1 Some Points on Negotiation Skills

Skill	*Action*	*Purpose*
Label Your Intentions	Signal your intentions and make the target clear	To direct the negotiation in small steps towards your goal
Test Understanding	Use questions to ensure that the other party has the same understanding	The 'feedback loop' discussed in the communication section is also important in reaching and confirming an agreement within negotiations
Summarise Agreements	Recap, agree and record any concessions or conclusions reached	Confirming and recording small steps agreed upon within the overall negotiation can both help maintain positive momentum and help reach the final resolution
Use a Quality Argument	Focus on the strong points in your argument, don't present a long list of weak points	Your position will usually become weaker if you present points that are easily knocked back. Know the strong points in your case and present them supported by relevant detail – don't dilute a good argument
Use Questions	Where possible ask questions where you think the opposing argument is weak	By asking for additional detail or thoughts the opposing view may become weaker or the other party may talk themselves out of believing in it
Watch the Clock	Try to establish time limits and stick to them	It is easy for certain points within a negotiation to be forgotten or unconcluded at the end meaning another meeting may be needed
Think Outside the Box	Try to find a resolution that is good for all parties – the win-win	If a negotiation is with a party you expect to work with again, then this is particularly important. If instead you try to achieve a situation where you win and the other party loses, this could make future collaborations difficult – or more expensive! If on the other hand you know you will not be working with the other party again, then this is less important
Ask Questions	Use the questions to persuade and to control the conversation	Use questions to reveal the strength of your argument rather than to force your argument on others. Questions can also be used to focus the conversation in areas that you want to discuss, and also to avoid sounding negative. Rather than saying something won't work, consider questioning to allow the other party to see the weaknesses in their ideas, such as 'can you tell me how that could work here?'

One factor not included in Table 6.1 is the need to understand your negotiating power and where this comes from. If one party has significantly more negotiating power than the other, then they should have the advantage in the negotiation if they recognise their power and use it well. Table 6.2 below lists potential sources of negotiating power and describes what each means.

Having reviewed the source of any negotiating power you have, then before the negotiation begins, the bargaining tactics that will be used should be considered. Should the deal be broken down into small packages or combined into one? By breaking it down into small packages it may be possible to negotiate a higher price for example. By combining all elements of the deal into one it may be possible to make the deal appear more attractive.

Table 6.2 Some Sources of Negotiating Power

Source of Negotiating Power	*Explanation*
Need/Commitment	which party needs the negotiation more? The more you need to reach a conclusion then the more power the other side will have
Alternatives & Options	What options does each party have if agreement is not reached? The better your BATNA (best alternative to a negotiated agreement) is, the more power you have
Time/Deadline	If the other party has a tight deadline that they are working to and you don't, then you have more power
Relationships	If your organisation is of strategic importance to the other company and this is not reciprocated, then you have more power
Investment	The more effort or resources one party has invested, the more committed they will be to reaching an agreement and this will give more power to the other side
Credibility	If you have greater expertise or credibility in terms of what is being negotiated, the greater the power you will hold
Information/ Knowledge	Increased levels of information or knowledge will lead to greater credibility and hence greater power
Material Resources	If the negotiation concerns material resources, whoever controls the resources will have greater power
Skills	Whichever party has the best grasp of negotiating skills, as outlined in Table 6.1, will have a source of power
Perception	The more power the other party thinks you have, the more power you actually have
Elegant Solutions	One way to influence the other side in a negotiation is to provide a good solution to the problem. The more complex the problem is the more influential an elegant answer is
Authority	Negotiations sometimes take place between parties where one has greater authority than the other and this carries over into the negotiation
Coalitions	Negotiations are often between more than two parties. In these cases two parties can form coalitions thereby increasing their power

When negotiating never be afraid of saying nothing! A silence will be filled, often by the other party offering further detail or concessions if you wait. Know which parts of the deal are tradable and which are not. Don't feel obliged to react to pressure, don't give concessions without costing them out. Lastly know your BATNA (see Table 6.2), and at what point you should walk away from the negotiation.

6.2.4 *Conflict Management*

As a project progresses there will inevitably be sources of conflict which need to be managed. The APM[1] define conflict management as the ability to identify, address and resolve issues between individuals or groups. The reason that this is necessary is that, left unresolved, the conflict is likely to have a direct impact on the duration of the project, its cost or the quality of the end product. Assuming the conflict to have a negative impact on the project, the sooner this is addressed the better.

Is all conflict negative though? This chapter includes the subject of teamwork and how teams are formed in a later section. Teamwork is one area where conflict can be positive and can improve project delivery. Consider the case of a team as it begins forms. Perhaps none of the team members are clear of their roles, and so some assumptions may be made, leading to some overlap and feelings of having your 'toes stepped on' by others. This can easily develop into conflict, but it could be positive, as long as it doesn't get out of hand as it will lead to greater definition and understanding of roles helping in the efficient delivery of the project. Similarly, a project involving cutting edge technology may spark dispute and conflict linked to how a particular function or capability should be delivered. This type of conflict can help to find the best solution for the project and indeed be an important step in the project's successful delivery. Constructive conflict can help to bring concerns into the open, to reduce the level of misunderstandings, to raise otherwise unexpressed viewpoints and, perhaps most crucially, reduce uncertainty within the project.

In order to manage conflict, an organised approach needs to be taken, such as the five steps shown in Figure 6.7 below.

Poor levels of project communication, as discussed in Section 6.2.2 above, are a potential source of conflict. Low levels of trust, poor cooperation within the team, badly designed project procedures, and even simple personality clashes are additional sources of conflict which the project manager needs to be aware of and ready to address. If resources are allocated to project teams centrally then this can be an additional source of conflict and can be driven by different views, or misunderstandings of an organisations priorities, values and approaches. Conflict can be generated from inside the project team or from outside it and can be divided into three main categories, project goals, project personnel and project administration. These categories and the potential sources of conflict within each category are summarised in Figure 6.8 below.

If some level and types of conflict can be constructive then the project manager's job becomes twofold. On the one hand the project manager needs to ensure that any constructive conflict

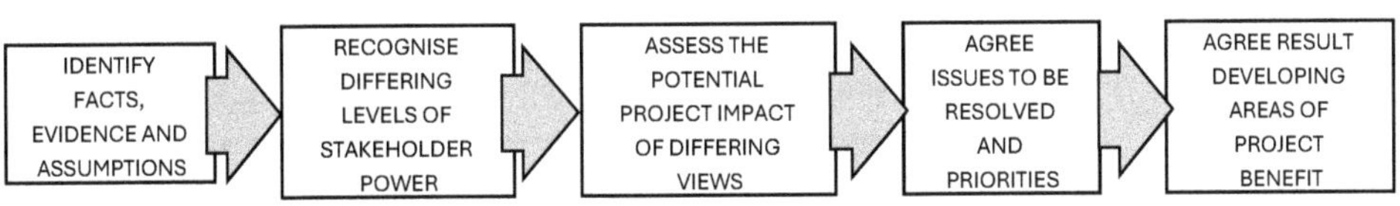

Figure 6.7 The five steps of conflict management within a project.

remains constructive and on the other they must focus on and address the negative, unnecessary conflict. The diagram shown at Figure 6.8 helps to focus on this. Clear communication of the project goals backed by strong sponsor support can help reduce the risk of conflict. Understanding the project personnel and their strengths, in terms of personalities and working styles, will help the project manager to structure the team effectively. Clarity around the project administration tasks will add clarity and help to reduce conflict.

When managing conflict within a project, it is worth noting that individuals can have different approaches to conflict which will help define how they handle the conflict. These are described in the well-known Thomas Kilman Conflict Model which is included in the APM's BoK7.[1] An interpretation of the model itself is shown in Figure 6.9.

Each of the categories shown in Figure 6.9, most of which will have a negative influence on the project, will require different approaches in terms of how the conflict is managed.

An 'Avoiding' approach where both the cooperativeness and assertiveness desires are low runs the risk that the issue causing the conflict will not be tackled meaning that the level of

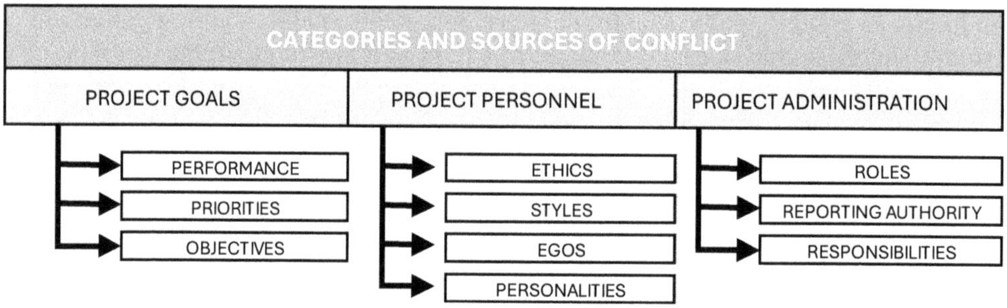

Figure 6.8 Categories and sources of project conflict.

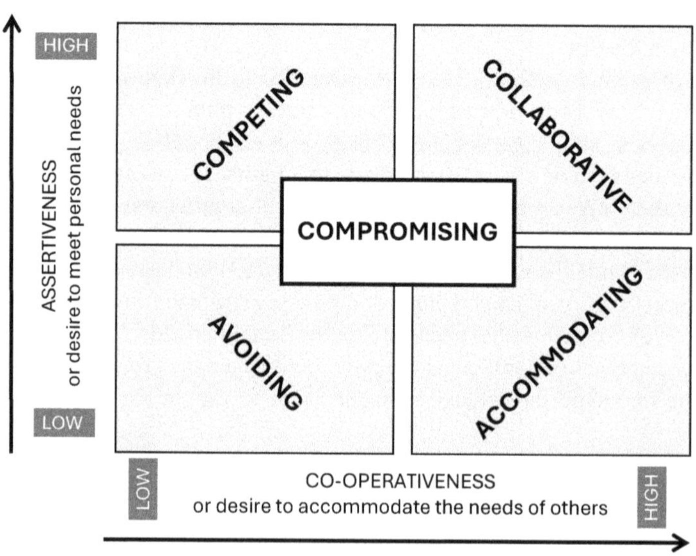

Figure 6.9 How the desire to meet personal needs and the needs of others impacts conflict resolution.

conflict will potentially increase. An 'Accommodating' approach will mean that the overarching aim of the parties is to keep the peace and create good will. This may seem a good approach, but the danger for the project is that it can fall victim to decisions made in order to keep the peace rather than to advance the interests of the project. A 'Compromising' approach is exactly that and again whilst it may allow the project to proceed, the decision reached may not serve the project in the best way. The fourth way in which conflict can be addressed is through a 'Competing' method. This will be detrimental to the project where it becomes a test of strength and both sides want to assert their personal needs whilst having little thought about accommodating the needs of others. If however, individuals value their own needs highly, but also wish to accommodate the needs of others, then the 'Competing' approach to the conflict will become 'Collaborative' and take both sets of views into account. This will mean that the best resolution for the project in terms of its impact on project efficiency and effectiveness is likely to emerge and to become adopted.

6.2.5 Stakeholder Management

One of the questions often asked in Project Management is 'what is project success?' An answer to this is that it is, and can only be, defined by the project's stakeholders, those impacted and affected by the project's implementation. The aims of the project are encapsulated in the business case (see Chapter 2) so in response to the definition of project success it must be asked 'who are the stakeholders and how do they interpret the business case?'

In addition to being impacted by the project, stakeholders also have the ability to affect one or more areas of the project. This is a wide definition and can draw from both inside and outside the project. Consider a construction project. Internal stakeholders will include individuals or groups such as the project team, the client and the suppliers and contractors. The make-up of external stakeholders will be even more varied and can include the planning authorities, local communities, environmentalists, local government, and the media amongst others. The impact of not managing local residents as stakeholders is evidenced in the project delays caused by planning appeals and even protests. It is not only time that can be impacted, but costs too can increase due to stakeholder actions. Delays in projects will inevitably increase costs, consider for example, the case of the A34 Winchester-Preston Trunk Road, more commonly known as the Newbury Bypass, built in the late 1990s. The protests by environmental groups not only delayed the construction of the road and caused an increase in project cost, but also added £25m, roughly a fifth of the total cost, to the project just in terms of the money spent on the private security required[5] which at the outset would not have been factored in.

There are of course many other examples of the actions of stakeholders having an impact on the time, cost and quality parameters of a project all of which only go to underline the importance of understanding and managing stakeholders.

The management of stakeholders has much in common with the preceding sections of this chapter, particularly those focused on communication and negotiation. It is necessary to understand stakeholders view of the project and to map their levels of engagement with it, in order to know the relative levels of effort that the project manager should dedicate to each stakeholder group. Stakeholder analysis can be performed using a process not too dissimilar to that outlined in Chapter 9 for the analysis of project risks. Stakeholders are in a way a particular class of project risk and mismanagement of them can be disastrous for the project.

One key stakeholder characteristic is their power and hence how big an impact, both positive or negative, they could have on the project. This can be mapped against the stakeholder's

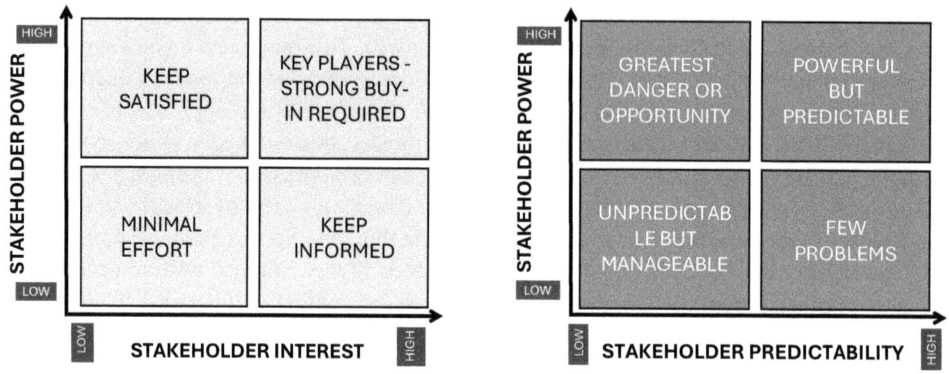

Figure 6.10 Two different ways of approaching stakeholder analysis.

interest in the project and the resulting plot used to describe how the project manager should approach their communication strategy with each of the main classifications. Equally stakeholder power can be mapped against stakeholder predictability to describe the amount of time the project manager should invest in understanding the individual stakeholder, their motives and the predictability of each. Both of these analysis frameworks are shown in Figure 6.10.

Once stakeholders have been analysed, the project manager can consider how they should best choose to engage with the stakeholders. The level of engagement could be anywhere on the spectrum of notify/inform/consult/Involve/collaborate and matters such as the method of engaging and the anticipated effect should be noted against each stakeholder, with the aim of maximising the benefits delivered by the project.

6.2.6 Teamwork

There is an old African proverb which says, 'if you want to go quickly, go alone, if you want to go far, go together'. In this sense projects that deliver significant benefits and outputs are like long journeys, they need a team of people to complete them.

Managing that team becomes one of the key responsibilities of the project manager.

The Katzenbach and Smith model of team effectiveness over time is well known.[6] It relates how any team, such as a project team, should develop over time moving from a 'working group' to as 'high-performance team'.

In terms therefore of team building, the project manager's key responsibility is to accelerate the team's progression through each stage of the team forming cycle shown in Figure 6.11, until they reach that of a high-performance team thereby becoming a highly effective, high impact team that can maximise the delivery of the project.

The uncertainty with which teams are faced when delivering projects can be off-set through the communication of clear goals and team structure including definition describing how the team will work together. This will make each individual's role clear and detail what is expected of them. The team needs this certainty of roles, together with clear communication of the projects purpose and direction.

It has been said already that teams can experience an element of conflict at the outset when team members are establishing their working relationships. The team themselves are important stakeholders in the project so a plan to engage with them and to communicate effectively needs to be considered. The team may have been selected directly by the project manager or may have been assigned, almost at random to the project. Either way, it is important to consider

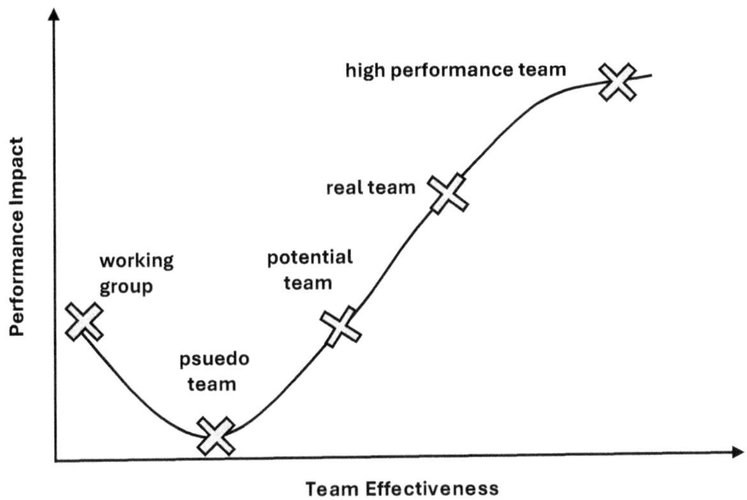

Figure 6.11 The Katzenbach and Smith model of team effectiveness.

the middle section of the diagram at Figure 6.8 'Project Personnel'. Can any conflicts between project personnel be anticipated? Can any tension between the ethics, styles, egos or personalities be identified as potentially problematic? If identified, then mitigation is likely to come from one of two sources. Either the establishment, adaption or re-enforcement of the right-hand-side elements of Figure 6.8, (roles, reporting authorities and responsibilities), or by communicating, with increased clarity the aims and objectives and the organisational structure of the project.

6.3 Project Review: The Impact of Soft Issues on the Project: The Fire Control Project

6.3.1 *The Project Background*

The Fire Control Project is a great example of a project that was impacted by multiple issues, many of them linked to the soft skills which have been discussed in this chapter. The view of Fire Control as a troubled project is supported by the House of Commons Committee of Public Accounts who state in their report[7] that the project was flawed from the outset.

The Fire Control Project, launched in 2004, was ambitious and aimed to improve the resilience, efficiency and technology of the national Fire Brigades. It centred around three key deliverables. These were,

- the replacement of forty-six local Fire and Rescue Services control rooms throughout England with nine purpose-built regional control centres,
- the provision of one common national computer system, and
- to help drive forward the business change within the Services, helping to embed new policies and procedures.

These aims required full engagement with the Services and clearly there was a need for stakeholders, particularly end users, to both help the design and support the acceptance testing of the new IT systems together with the drafting and embedding of the new policies and procedures.

The project, however, was cancelled in 2010 after a series of delays. None of the original objectives had been achieved and a Figure of at least £469 million had been spent.[7]

In July of 2011, the NAO[8] reported that at the time the project was cancelled, there was a five-year anticipated delay, and that the by the time of their latest forecast of total project costs, the original sum of £120m in 2004, had risen to £635m. Some monies were recoverable, but overall, an estimated £469m was lost as a result of the project's failure to deliver.

6.3.2 The Project Issue

The project failed to deliver the proposed aims and was never able to convince the local Fire Services of its benefits.

A House of Commons[7] report identified that the project was begun too quickly, without applying basic project approval checks and balance. Key amongst these was the failure to consult with the forty-six regional Fire Services who were reluctant to change the way they worked. Rather than discuss project issues with this key stakeholder group and to seek some form of negotiated agreement as to the way forward and how the project should be implemented, the project excluded the Fire Services from decisions relating to the design of the regional control centre buildings and the proposed IT solution. In order to gain the planned project benefits the Fire Services, as the end users, were the very people who needed to be brought on board and to contribute in order to ensure benefit delivery. Without their acceptance and knowledge, the desired project benefits were put at risk.

There was pressure to deliver the project quickly, driven by the desire to ensure that there was a well co-ordinated response to the potential national threats of terrorist attacks, flooding or railway crashes. As a result, the project took major decisions before a business case, project strategy or a procurement strategy had been formulated or discussed and agreed with the Fire Services. Both project costs and risks were underestimated whilst the project started with an over optimistic view of the deliverability of the proposed national IT system in particular. A primary cause of this failure was a lack of understanding of requirements brought about by a lack of engagement with the end users.

There were challenges too in respect of project leadership and teamwork. Consultants comprised more than half of the project's management team but appeared not to be directly managed on the client side. Roles and responsibilities were not clear, and the project suffered from high levels of turnover within its senior management. High levels of staff turnover can often be a sign of a failing project (see Chapter 14), and it is noted in the report[7] that there was little accountability for project failure apportioned to the senior managers. This was deemed a key factor in what was described as 'an extraordinary failure of leadership'. It contributed to a poorly negotiated IT contract, awarded to a company with no previous direct experience of working with the emergency services. It was also difficult to hold this contractor to account due, in part, to the way in which the contract was written without early delivery milestones or checkpoints being included within it.

The NAO report[8] states that the lack of clarity within the project, evidenced by a lack of 'communication and engagement' with stakeholders led to 'concerns about its rationale and purpose'. It was unclear as to how the proposed changes would increase the efficiency of the Fire Services. Lack of clarity around the project benefits reduced levels of support for the project particularly given the lack of consultation with the local Fire Brigades, the need for which had been specifically recognised in an earlier report.

The Fire Brigades Union website[9] bemoans the waste of resources brought about by the project, in particular that, as of August 2017, four out of the nine buildings designed to be used as regional centres stood empty and unused, seven years after the project was cancelled. The cost of maintaining the buildings over the years has been considerable. As late as 2022, two sites remained unsold. The sites had been built in 2007[10] and the nature of the project's financial structure meant that the buildings were funded via a 'buy and lease back' approach, where the assets, in this case the Fire Control regional centres, were bought by a third party and leased back to the regional fire centres, over a period of 20–25 years. Lease payments were committed to and therefore had to be made for the duration of the lease period irrespective of whether the centres were used or not. In the case of the Castle Donington control centre,[10] the lease runs out in 2033 at a continuing cost of almost £2m per year. The building has laid empty since its completion in 2007 despite efforts to find new tenants.

6.3.3 How Could Better Use of Project Soft Skills Have Helped?

Witness statements given to the House of Commons Committee of Public Accounts on the failure of the Fire Control Project (2010–2012) put forward the view that the government had moved 'ahead on projects on the basis of its assessment of the issues and (did) not sufficiently connect with the local players.' This points to a lack of engagement and a lack of ownership by the Fire Brigades leading to a flawed project design. This could have been avoided through greater consultation with the brigades, greater negotiation, communication and stakeholder engagement. Amyas Morse, head of the NAO at the time of the report, highlighted a key issue 'that they (the Fire Brigades) were being given residual liabilities (that) they had not had an opportunity to represent themselves and negotiate'. Clearly this approach should have been seen as likely to cause rancour, let alone failing to make the most of a key input into the design of an efficient system. This would have been largely addressed through application of the stakeholder analysis process outlined in Figure 6.10 and the consideration and identification of sources of project conflict outlined in Figure 6.8. Early identification of possible issues for the project in these areas could then have been used to feed into the development of a communication plan such as that shown in Figure 6.5. Work would have been required and it might not have been straight forward, but one effect of this would hopefully have been to improve the teamwork within the project. The brigades were a key part of the project's benefit delivery, they were the people using the project's outputs, so needed to be part of the project team. Non-engagement with them ran the risk of stagnating team development in the 'working group' area, at best, of the model of team effectiveness shown at Figure 6.11. The Project Manager's responsibility is to develop the team as quickly as possible into a 'high-performance team'. In light of the comments made in the reports about project leadership, it would appear that this aspect of project management was sadly missing.

6.3.4 Conclusions on the Fire Control Project

So much seemed to go awry with the Fire Control Project that it was reported[7] as one of the worst cases of project failure that the House of Commons Committee of Public Accounts had seen for many years. Engagement with the Fire Brigades was largely absent, leading to a lack of buy-in and a lack of understanding of the risks the project was, to an extent, creating for itself. Basic project processes were not followed. Project leadership failed to direct the project successfully in a number of areas, particularly in the awarding and management of the IT contract. It is interesting to see that following the cancellation of the project, bids to improve resilience, efficiency and interoperability were invited from each of the Fire and Rescue Services, thereby ensuring a degree of buy-in.

In addition to the issues in the preceding paragraph, The House of Commons report[7] also concluded that the lack of expertise in delivering the project meant that there was an over reliance on consultants whose performance was not effectively managed, and focus was thereby lost on what was being delivered. The interoperability between the three key parts of the project, outlined in Section 6.3.1, was absent and this was confounded by a lack of clarity over project roles and responsibilities identified. This lack of clarity meant that despite the scale of project failure and the resultant financial waste, no one was held accountable. The project manager should have been facilitating and confirming the definition of the roles and responsibilities, the fact that they weren't held accountable perhaps suggests that the role of project manager was not well established and did not carry the weight that it should have, either internally to the project or externally.

Section 6.3.1 begins by saying that the Fire Control Project is a great example of a project that was impacted by multiple issues. The review above illustrates how if soft skills are not employed wisely, then a complex project can quickly become impossible to deliver and this invariably will begin to impact the project in other areas.

6.4 Soft Skills Before the Birth of Modern Project Management*

(* In this section, the research relating to the construction of HMS Barham 1912–15 mentioned in the introduction is drawn upon to help illustrate how the subject of this chapter has developed.)

Evidence for the use of soft skills within projects can be difficult to find. Whilst leadership, negotiation, communication, conflict management, stakeholder management and teamwork are present to a greater or lesser extent in all projects, the ways in which they are delivered can be less obvious given their less tangible nature.

Communication is evidenced in the case of the WW1 construction of HMS Barham through a number of letters that have survived detailing such aspects as the authorisation of ships drawings and requests for increased labour. Similarly signed contracts indicate the presence of negotiations and stakeholder management is evidenced through the regular communications with the Admiralty as the ships build progressed. Whilst there is evidence of all these elements, what is missing is evidence of the processes behind these actions. No evidence of stakeholder mapping was found or of a communications plan, nor was any evidence found of the other soft skills. It is suspected that this is due to incomplete records surviving in the archives after 100 years, but that these processes, or similar, would have been in place.

The construction of ships within HM Dockyards at the beginning of the 20th Century was well governed, and this is best illustrated by a 1909 publication (still in use at the time of HMS Barham's construction), with the fabulous title of '1909 Instructions to the Professional Officers of His Majesty's Dockyards at Home in Matters Relating to the Professional Duties of their Departments Together with Directions Relative to Building, Fitting and Completing Ships, Defects, Alterations and Additional Fittings in Ships and Estimates and Expense of Work.'[11] (Illustrated previously at Photo 5.1) The instructions cover the duties of those employed, and the nature of documents such as accounts and ships' drawings that needed to be provided to the Admiralty, and it is certainly a comprehensive communication document for all those employed in the Dockyard.

The book illustrates the Admiralty's approach. Doubtless private companies would have needed to comply with Admiralty demands just as much as HM Dockyards would have done, yet the archival search did not provide any records reflecting this.

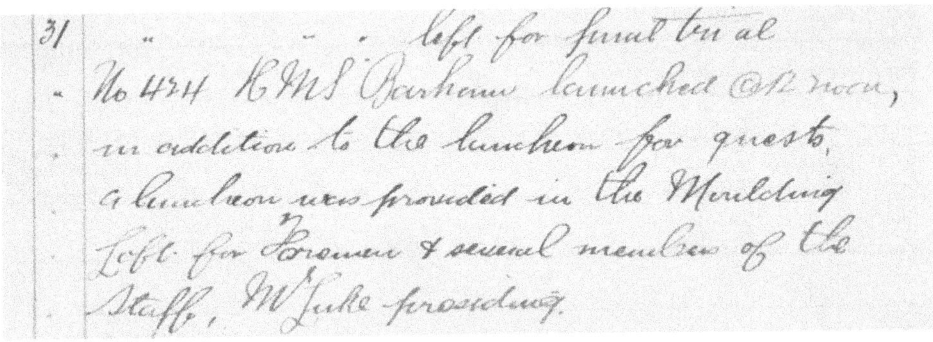

31 " - . left for final trial
No 424 HMS Barham launched @12 noon,
in addition to the luncheon for guests,
a luncheon was provided in the Moulding
Loft for Foremen & several members of the
Staff, Mr Luke presiding.

Photo 6.1 An extract from the John Brown Dockyard's 1914 Shipyard Dairy (Reads: No 424 HMS Barham launched at 12 noon, in addition to the luncheon for guests, a luncheon was provided in the Moulding Loft for Foremen and several members of the staff, Mr Luke presiding). By permission of the National Records of Scotland and University of Glasgow Archives & Special Collections, Records of Upper Clyde Shipbuilders Ltd, Clydebank Division collection, GB248 UCS 1/10/2.

Which records survive and are kept is determined at least in part by their perceived importance. It is also governed by luck and the ability of the archive to survive over time. The records of the John Brown shipyard, where HMS Barham was constructed, were thought, for many years, to have been destroyed during WWII bombing but were found and partially rescued in the 1960s. This could be the reason why minimal records of soft Project Management processes were discovered during the research.

One record that did survive however, is evidence of the celebration of the team's success through an 'end of project party', something which is familiar to projects of today. Photo 6.1 shows an entry from the Shipyard's Diary. The entry on 31st December 1914, marks HMS Barham's launch stating that 'in addition to the luncheon for guests, a luncheon was provided in the Moulding Loft for foreman and several members of the staff'. Whilst this does not include the full project team, along with the other documents discussed, it does suggest that some of the softer skills elements were present, reflecting those used in today's projects.

References

1 APM. (2019). *APM body of knowledge* (7th ed.). Association for Project Management.

2 Project Management Institute. (2021). *The standard for project management and a guide to the project management body of knowledge* (7th ed.). Project Management Institute, Inc.

3 Ohara, S. (2017). *P2M – A guidebook of project & program management for enterprise innovation* (3rd ed.). Project Management Association of Japan (PMAJ).

4 Chartered Institute of Building. (2009). Leadership in the construction industry. https://www.ciob.org/sites/default/files/CIOB%20research%20-%20Leadership%20in%20the%20Construction%20Industry%202008.pdf [Accessed 8 August 2024].

5 The BBC. (2016). Did the Newbury bypass tree-huggers change anything? https://www.bbc.co.uk/news/uk-england-berkshire-35132815 [Accessed 10 August 2024].

6 Katzenbach, J.R., & Smith, D.K. (2015). *The wisdom of teams: Creating the high-performance organization.* Harvard Business Review Press.

7 House of Commons Committee of Public Accounts. (2011). The failure of the FiReControl project. https://publications.parliament.uk/pa/cm201012/cmselect/cmpubacc/1397/1397.pdf [Accessed 6 June 2024].

8 The Comptroller and Auditor General. (November 2011). *The failure of the FiReControl project* [online]. The House of Commons. https://www.nao.org.uk/wp-content/uploads/2011/07/10121272.pdf [Accessed 4 June 2024].
9 The Fire Brigades Union. (November 2017). *Exposing the shambolic regional fire control project.* https://www.fbu.org.uk/blog/exposing-shambolic-regional-fire-control-project [Accessed 4 June 2024].
10 Kreft, H. (2022, February 16) The fire control centre that never opened has cost taxpayers £20m. https://www.staffordshire-live.co.uk/news/local-news/fire-control-centre-never-opened-6656171 [Accessed 7 June 2024].
11 The National Archives of the UK (TNA) ADM 95 107 ADM.

Chapter 7

Estimating

Estimating is often thought of as just an initial part of scheduling and budgeting, the focus of Chapters 8 and 10, respectively. However, its influence is much wider than this. Throughout the delivery of a project, the project manager will need to estimate a wide variety of elements such as the degree of risk that a project may be subject to along with the size of those risks. The potential influence of a stakeholder will need estimating too, in fact any unknown that could impact on the time, cost or quality of the project outputs will need to be estimated if it is to be managed.

7.1 The Purpose of Estimating

Estimating, within the project environment, is the process of establishing expected values or anticipated values for variables such as cost and task durations. Without a set of estimates base lines cannot be established and the project manager will have nothing against which to control and monitor the project (see Chapter 11). The PMI[1] define an estimate as 'a quantitative assessment of the likely amount or outcome of a variable, such as project costs, resources, effort, or durations.' In doing so, they say an estimate must have a numeric value and link it to activities such as budgeting and scheduling. The APM[2] also consider an estimate to be a forecast of time or cost, which is also consistent with the approach of P2M.

One of the purposes of cost estimation has been defined (P2M[3]) as being something that is required in order to explore the concept and viability of the project in its early phases to enable a budget to be agreed so as to facilitate the raising of investment capital for a project. This in turn would support the competitive bidding process and as noted above, establish a baseline budget against which project delivery can be monitored. Estimation within a project is not simply just the numerical values produced, but it is the process that provides the tools or methods that can be used to judge the potential impact of various elements of the project on the ultimate time, cost, and quality of its outputs.

P2M[3] states that, an accurate estimation is needed from the early stages of the project if additional costs are to be minimised for the project. This is of course true (over estimates become self-fulfilling whilst underestimates introduce friction to the project), but it raises the question of how an accurate estimate can be established and what are the approaches available to the project manager in order to do this?

DOI: 10.1201/9781003470748-8

7.2 How Can the Project Manager Approach the Estimating Task?

As discussed in Section 7.1, there will be many things that a project manager needs to estimate within the delivery of a project. This inevitably links back to the time, cost and quality parameters discussed in Chapter 2. The project manager will need to ask themselves several questions in order to choose the best approach to the estimating process, such as what? when? who? how? and even why? Estimating is important since all project resources are finite which will impact on the choice of subcontractors, the scope of the project, the cost benefit analysis and hence whether or not the project should even be started. With such key decisions being made, it is vital that enough consideration and time is given to the estimating process to ensure that it has the required level of accuracy for planning purposes and to allow it to form the basis of a reliable baseline against which the project can be monitored and controlled. Once the initial estimation process is completed it will be possible for the project manager to know how long the project is expected to take (the time) how much will need to be spent (the cost) and the approach that the project can afford to take (the quality).

An estimate can be made for the project as a whole, and this will also enable the project's profit to be calculated, profit (or loss) being the gap between the project cost and project revenue, as shown in Figure 7.1 below.

As presented, this assumes that the revenue is received throughout the project's lifecycle – the gradient of the Planned Revenue curve is always positive. This will, of course, depend upon the nature of the project. Remuneration may not be received until once the project is completed, as in the case of the sale of a new building. Alternatively, it may only be received once a new process has been delivered and is operating within the in-house environment, delivering costs savings, rather than enabling direct revenue to be received. Regardless of which is true, the net benefits will need to be estimated as part of the business case to confirm the project's viability.

7.2.1 *Challenges Faced by the Estimating Process*

Estimating is not always a simple case of just producing an expected cost or duration for a project task.

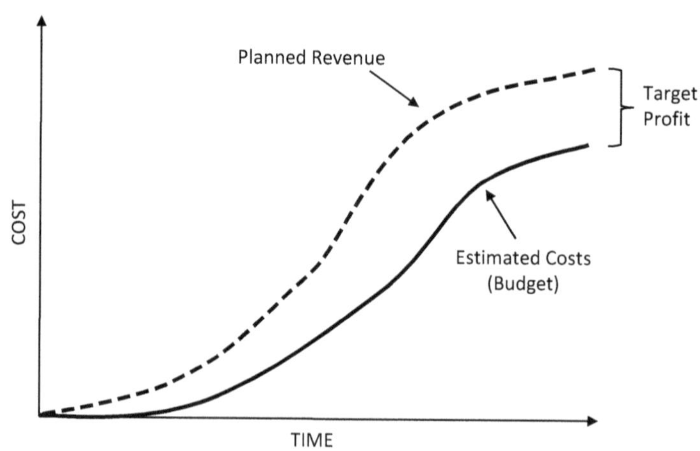

Figure 7.1 The estimation of a project's profit.

Estimating is a through life task for the project manager, but a key part of this is undertaken in the early stages of the project. At this stage, there can be many unknowns such as the level of hidden risks or the potential volatility of the environment into which the project delivers, hence the exact scope of the project deliverables required can be uncertain.

The relatively poor level of project knowledge at this early project stage can be the cause of poor estimates due to the teams less than complete understanding of the project. This lack of precise definition can also mean that there is a lack of specialist quality input in terms of expertise or data into the early estimates. The team should be aware of psychological factors when sourcing estimating data. It is important to think about how the estimate is being requested. If a senior manager within the organisation asks for an estimate, how is the question phrased? Due to optimism bias or political pressure, if the question is phrased as 'could this task be completed by the end of the month?' rather than 'how long do you think this task will take?' a very different answer could be obtained. Additionally, if, for instance a time estimate is being sort, then what are the assumptions of the person giving the estimate? One key aspect to note in respect of an estimate of time given for a task to complete is whether it is a 'durational' estimate or an 'effort based' estimate. The difference between the basis of these two estimates is that the first implies that the task could be completed within the next month but does not specify the actual hours spent directly on the task involved. The second, the 'effort based' estimate, states how long the task will take based on working full time on that task and that task alone, but does not say when this work could be done. It is important for the project manager to be aware of the difference as this can help if a project is running late. Knowing that you have a scheduled a task using a 'durational' estimate rather than an 'effort based' one means that there is potential to bring the completion of the task forward within the 'durational' estimate window given.

Cultural differences can influence an estimators' approach to issues such as risk and this can be reflected in the estimates received. Cultural differences can exist within an organisation, perhaps between departments, as well as the more obvious example of between the varying international project sites of a multi-national project.

The stage of the lifecycle in which the project finds itself is a well-known cause of variation of accuracy in estimating, and it gives rise to the term 'estimating funnel'.

Figure 7.2 demonstrates the 'estimating funnel' which illustrates that as the project proceeds, the variation of estimates from the actual figures will diminish (Figures showing the indicative accuracy of the estimate are purely illustrative, indicating the relative accuracy of estimates over time).

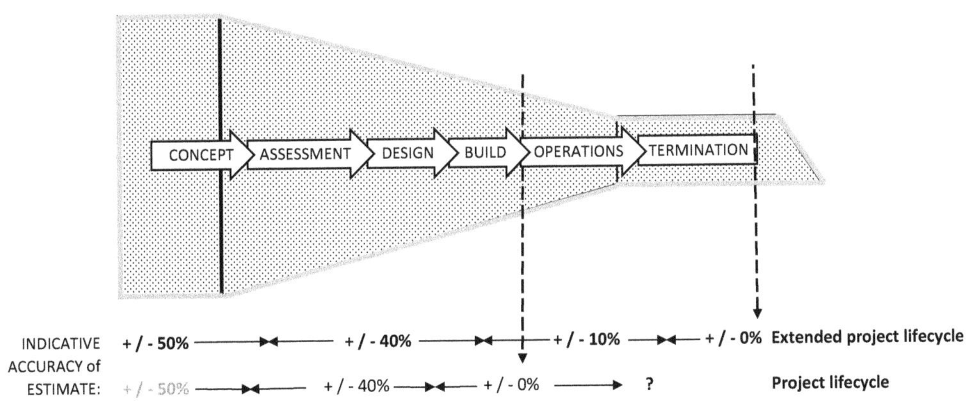

Figure 7.2 The estimating funnel.

The reason for this is that as the project progresses, more of the costs and time estimates become actuals, thereby reducing the room for inaccuracy within the estimate. At the same time, the number of unknowns reduces as they become recognised and project risks either occur or pass the moment in which they could occur, further adding to the certainty of out-turn costs and thereby reducing the estimate band still further as the project advances. The diagram shown at Figure 7.2 illustrates both a project lifecycle and an extended project lifecycle. The difference between the two is the question of whether or not the operational and disposal phases are included in the lifecycle and hence the need to provide estimates for this section or not (See Chapter 4). The question mark shown in relation to the 'lifecycle' and operational section shows that these costs will not have been borne nor estimated by the project itself, but will be paid by the operator using the project outputs.

It has been said[17] that project underestimates lead to disaster, whilst overestimates becoming self-fulfilling prophecies. This then leads to the conclusion that the only useful estimate is an accurate estimate! Some reasons for poor estimates are noted above, perhaps though the final reason for poor project estimates is a lack of experience or awareness of estimating skills themselves. There are four ways in which projects generally produce estimates, these are explored in the following section.

7.2.2 *Methods of Estimating*

It has already been said that project estimates focus on the evaluation of costs and durations within the project. If costs are the emphasis of the estimate, then the range of estimates within the project will cover items such as the materials required, and the labour or expertise required to assemble the output. These can be thought of as being internal to the project, things over which the project team will have an element of choice or control. Estimates are also required for components external to the project such as company overheads or even taxation, inflation or, if it is a project with an international aspect, exchange rate fluctuations, over which the project manager will have little control.

The estimating methods themselves can be divided into two different genres, qualitative, where the estimate is based on the thoughts of an individual or group, and quantitative where the approach is very much more numeric and potentially more involved. Both can be top down or bottom-up, defined in relation to the Work Breakdown structure discussed in Chapter 8. Top down simply means that the process begins with the top line in the project breakdown, and then moves downwards through the project's outputs. This tends to be a less granular approach than the bottom-up technique which focuses more on each and every task at the lowest level of the Work Breakdown structure before moving up through the levels within the Work Breakdown structure and combining the individual estimates to give an overall assessment of project cost.

Whether the method chosen is a top down or bottom-up approach, the estimator will need to select either a qualitative or quantitative method. There are two principal qualitative methods known as either a subjective or comparative process, whilst the quantitative process dictates either a parametric or analytical method. A review of each in turn will demonstrate the key differences between the methods.

7.2.2.1 The Subjective Estimating Process

A subjective estimate is based upon the experience and expertise of the person giving the estimate. How the individual providing the estimate is chosen is therefore of the utmost importance and should be carefully thought through. Obviously, the person who is asked should have prior knowledge of the variable being estimated, and the greater the degree of this knowledge or expertise, the more accurate the estimate provided should be. Often within a project the leader of certain workstreams will be a more senior member of staff who will have had experience of the task in hand multiple times. The foundation on which their estimate is built should therefore be solid and provide an informed basis for their assessment. There is however one potential danger than is often overlooked when sourcing estimates in this manner, and this is the character or situation of the individual providing the estimate.

In judging the character behind the estimate, what is important is to consider whether the individual is naturally pessimistic or optimistic, will they err on the side of caution or be gung-ho? The way and situation in which an estimate is requested can also impact the figure given. Suppose an expert, but junior, member of staff was asked to estimate how long a task might take by the organisation's managing director. If the question is posed as 'can you finish this by next Wednesday?' then the answer is quite likely to be yes, regardless of the reality.

The basis of any estimate was discussed above and irrespective of the character or situation of the estimator, it is crucially important to understand the foundation of the estimate provided, is the estimate 'effort' or 'durational' based? The ability to judge these aspects increases with experience, both in terms of the task being estimated and familiarity with the character and expertise of those providing the estimate. Consequently, the experience of those requesting the estimate is an important factor in making an accurate subjective estimate.

7.2.2.2 *The Comparative Estimating Process*

Comparative estimates require there to be some past history of estimating or of project delivery as they rely on the relation between past projects and future projects.

Part of the definition of a project is that they are unique, non-repetitive undertakings. As a result, any comparison cannot be a one hundred percent, direct evaluation and so it is important that the estimator fully understands both projects and the differences between them. In order to analyse projects in such a way, a detailed set of project records must be kept. For instance, if two towers are being compared, equal in every way except for their final height, then the project record for the first tower must be held revealing the height of this tower. Plans for the second tower will confirm the elevation of the new construction. If the proposed construction was twice the height of the first, it would be possible to conclude that twice the number facia bricks would be required at double the cost of purchase. This is a simplistic comparison, if for instance the total number of bricks required was to be calculated, then aspects such as the tapering or profile of the tower and the any additional depth of foundations required in order to bear the extra weight would need to also be calculated. This illustrates the need for detailed knowledge of both projects, past and present, in order to fully understand the differences and ultimately the impact of these differences on the estimates being made. This can be accomplished through careful collection of project records, something that it is useful to have support from a Project Office to enable comparisons to be made with as wide a range of projects as possible, and hence to be able to select one with a high degree of similarity.

Overall, the comparative estimating process can be more accurate than the subjective estimating approach, but it does require a basis to compare against, which can add both time and cost to the estimating procedure.

7.2.2.3 *The Parametric Estimating Process*

The third of the four estimating approaches is the parametric method. The dictionary definition of 'parametric' states that it relates to the parameters of something, to a mathematical or statistical variable. Within the project environment, a parametric estimate therefore relies on a set of statistics that can describe the item being estimated.

At its simplest, if a 100 ft wall took 10 days and cost £X to build, then using the parameter of length a 300ft wall would, all other things being equal, take 30 days and cost £3X to build. Similarly, a simple estimate can be made for the cost of building an aircraft wing simply by knowing its square area. This would enable the amount of steel needed for each side of the wing to be calculated, it would approximate the amount, or length, of wiring needed in the wing and consequently the number of man hours to install the wiring. These types of estimates can be used too when insurance is sold or quoted. In the case of house insurance, the insurer will ask a number of questions such as the property's postcode, the number of bedrooms, whether the insured is a smoker, an owner or a tenant etc. These questions will then enable the insurer, based on a large data base, to calculate the likelihood of the house being broken into, the property's size and hence rebuild costs should it burn down and the amount of care that the insured is likely to take over the property. All these, with the aid of a statistical equation, translate into the property's insurance premium.

In the case of a project, the parameters of each project task can be analysed and an estimate produced. This estimating technique is therefore a bottom-up process and different to the subjective and comparative approaches which are regarded as top down processes.

One of the advantages of the parametric approach is that, once the statistical information has been collected from previous cases, the estimate itself can be made quickly, as in the case of an insurance quote produced over the phone or online. It is likely to be more accurate than the two previous methods discussed, so when an accurate estimate is needed and there is a wealth of relevant, good quality, historical data available and it is possible to recognise, isolate and measure the various parameters of the project's tasks then a parametric estimating approach can be a good choice.

7.2.2.4 The Analytical Estimating Process

The analytical approach to estimating is just that. It is the most time consuming of the approaches to estimating and is very much a bottom-up process, addressing each of the project tasks in turn and in depth so as to build up an overall estimate for the project as a whole.

An analytical approach is heavily dependent on the Work Breakdown Structure as it takes as its basis the lowest level of the breakdown, it estimates the costs or durations for the components of these tasks, and then amalgamates these estimates into one figure for the project or workstream overall.

To do this accurately the project definition must be well developed, the tasks within the Work Breakdown Structure must be largely known and well understood. It is this high level of project granularity that facilitates the precision of the estimate and means that both project client and manager can see and understand where project costs are incurred. The increased accuracy enables better project control and improved tracking of actual costs against baselined budgets enabling the need for corrective action to be identified sooner. Any changes to the planned delivery can also be costed and the overall budget quickly adjusted due to the detailed nature of the information held.

The argument against using this type of approach however is the amount of time involved in producing the estimate, it requires detailed information about each task and it is resource intensive and it is therefore relatively costly.

7.2.2.5 The Choice of Estimating Method and 3 Point Estimating

Four estimation methods have been discussed, each with their own set of advantages and disadvantages. There will be times when one method is more suitable to use than others. For a quick estimate of an element of the project that is of relative insignificance, either in terms of cost or time, a subjective estimation may well be the best approach. Where the component part to be estimated is more complicated or more significant in terms of the project delivery, perhaps being on the critical path or being responsible for a large proportion of the project's cost or duration, then a more robust estimating technique such as an analytical method would offer benefit.

One potential reason that a project might want to take an analytical approach to estimating is to enhance the quality of thought that goes into the production of an estimate, the more thought used in producing the estimate, the more accurate an estimate can be expected to be, and the lower the risk of relying on a superficial, quickly produced single point estimate, would be.

An alternative way, rather than simply choosing the analytical method, to increase the level of understanding within a project is to consider using a 3-point estimating process. This is not a substitution for an analytical estimate, rather it can be used in tandem with any of the four estimation methods discussed above. What its usc provides is an increased level of understanding of the task being estimated and the potential risks associated with it, and hence it will provide a better quality of estimate.

3-point estimating is exactly what its name suggests. It is an estimate taken following the consideration of the task from three different points of view or perspectives. The process asks the estimator to consider what is, for example, the lowest duration a project task might take, producing an optimistic estimate where everything goes as well as it possibly could do. The second point to be estimated is taken from the most likely duration for the task, providing an estimate that can be regarded as there or there abouts more often than not. Finally, an estimate is provided from the third and final perspective, a pessimistic extreme where the worst (excluding Acts of God), is assumed to happen, thereby estimating what the maximum duration could be for the task. The process of contemplating these three points, or possibilities, encourages greater understanding of the task and can even help to identifying risks and opportunities associated with each task, as discussed in Chapter 9.

If using an IT based Monte Carlo simulation or similar, 1,000 or more estimates of the same task can be quickly made taking account of the impact of these three perspectives, the minimum, the most likely and the maximum durations. The shape of the distribution of these estimates around a central point can differ significantly, but in the absence of any additional project information, a normal distribution would be expected as shown in Figure 7.3 below, which shows that the most likely estimate is 8 (either days or £ks) and that the distribution either side of this is symmetrical.

3-point estimating can therefore take into account the impact of uncertainty, meaning the natural variation in a task. It can also be used to estimate the effect of risk on a project's delivery. If a risk was estimated as having a probability of 10% and an impact of 10 days in terms of duration if it were to occur, then given the 10% probability of occurrence, the presence of the risk would mean that the overall average estimate of the task's duration would increase by 10% of 10 days, or 1 day overall.

The horizontal axis of Figure 7.3 is a reminder of the need to employ an accurate estimating process. The project manager will need estimates for two key areas – the duration, or time taken, by a task and the cost of successfully delivering that task. The estimation therefore directly feeds

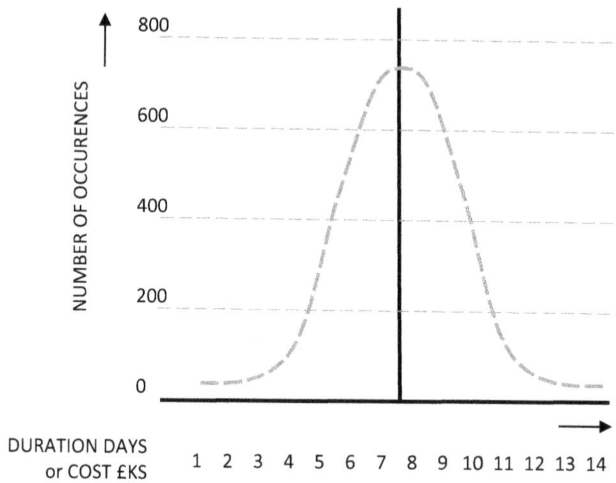

Figure 7.3 A normal distribution of values within a range of 3-point estimates.

into the project's schedule and the project's budget. Both these are key aspects of the time, cost quality triangle, and the baselined version of both will be agreed by the project manager with the project client and, as discussed in Chapter 2, be signed off with the client and, in turn, used by the project manager to monitor and control the project.

7.3 Project Review: The Impact of Estimation on the Project: High Speed 2

7.3.1 *The Project Background*

High Speed 2, or HS2 as it is commonly known, was seen as the 'strategic answer to join up Britain more effectively to meet the transport needs of the 21st Century',[4] it was described as 'the ambitious programme to create a new high-speed rail service from London to Manchester and Leeds, via Birmingham and the East Midlands.'[5] The project itself was formally announced in a press release from the Department of Transport in January 2012.[6]

The project was designed to be delivered in two phases. The first phase was to establish high-speed rail links between London and Birmingham, along 140 miles (225 km) of new track.[7] The second phase was to link London, via the first phase link to Birmingham, then onto Manchester, Leeds, the East Midlands and South Yorkshire.

It was stated in the Government's 2017 publication 'High Speed Two From Concept to Reality' that 'HS2 will become the new backbone of our national rail network. It will increase capacity on our congested railways and improve connections between our biggest cities and regions. It will support our Industrial Strategy, generating jobs, skills and economic growth to help us build an economy that works for all.'[8]

The list of expected benefits (consisted of faster journey times and increased capacity and for the first phase alone was calculated to deliver a Benefit to Cost ratio of 2.0), meaning that for every pound spent on the project by the Government, the Department of Transport (DoT) calculated, in 2011, that benefits worth two pounds would be received.[9] The report by the DoT concluded that over the next 60 years both phases of the project together would deliver a Benefit to Cost ratio of 2.6. The full business case for High Speed 2 Phase One was published in April 2020. It stated that HS2 offered huge potential for the whole nation which outweighed its costs. It would 'generate a vast increase in capacity, with thousands of extra seats, making life easier for travellers and better connecting our biggest cities. It had the potential to create highly skilled jobs, spread prosperity across the country and deliver a world class transport infrastructure.'[10]

7.3.2 The Project Issue

Crucially however, the HS2 Chairman's 2019 report[4] states that both the scale and complexity of the task were underestimated, leading to the budget and target schedule for the programme proving to be unrealistic, and to the potential benefits stated being inaccurate. All these elements, particularly the costs and benefits would have needed to be estimated and so the above report indicates that there were both difficulties and errors within the estimation process leading to both sides of the cost benefit analysis (discussed in Chapter 2) being mis-stated. Given that this was the case, then project funds would have been committed without the overall benefits of the project being accurately quantified or understood. The poor estimation and understanding of the cost of each task, led to costs continuing to spiral throughout the project, fuelling constant bad press and raising doubts about the project's viability.

If the cost estimates are reviewed between 2011 and 2020, then a clear trend is revealed, as shown below in Figure 7.4. Without further analysis, it is not possible to say exactly what caused the increases in cost. It could have been the result of increased accuracy or knowledge of the project parameters as the project moved from left to right along the estimating funnel (Figure 7.2) or it could be a result of things such as poor initial project design or inadequate estimating skills within the project. Each one of the increases shown in Figure 7.4 would have consumed management time in terms of managing expectations and balancing the pressure to reduce costs. Some of the early increases in the project's cost estimates, totalling around £12bn, were caused by the omission of costings for rolling stock and elements of the line's construction costs, suggesting that of the options above, an incomplete understanding of the project's component parts was at least a contributory factor. The pressures to moderate project costs and their estimates resulted in a reduction in forecast completion costs in 2017 when estimates relating to areas such as property acquisition, contractor efficiency, and overhead costs were revised downwards, but these unfortunately were not ultimately delivered.

In 2023 prices, Figure 7.4 shows that the estimated project cost rose from circa £50bn in 2011 to circa £130bn in 2020. This represents an increase of 260%.

With this degree of price increase it is unsurprising that the project was finally cancelled in October 2023. The Government issued a press release that revealed that they, as sponsor, had decided to re-redirect funds from the remainder of the HS2 project to 'build the daily connections that people depend on – unlocking potential, driving growth and transforming communities.'[11] This meant that rather than delivering the HS2 Phase 2 line between Birmingham and Manchester, the Government decided to 'take action to deliver greater frequency and quality of transport infrastructure across the whole country',[11] resulting in funds totalling £47.9bn were to be redirected into developing existing rail links chiefly in the North, Midlands and links between Manchester and Liverpool.

One of the reasons cited for the cancellation was the country's changing travel patterns post Covid meaning that although originally, every pound spent on HS2 was reported as delivering £2.30 worth of benefits back to the taxpayer, the figure at the time of a 2020 press release was restated to show a much reduced return in terms of benefit delivered, potentially as low as

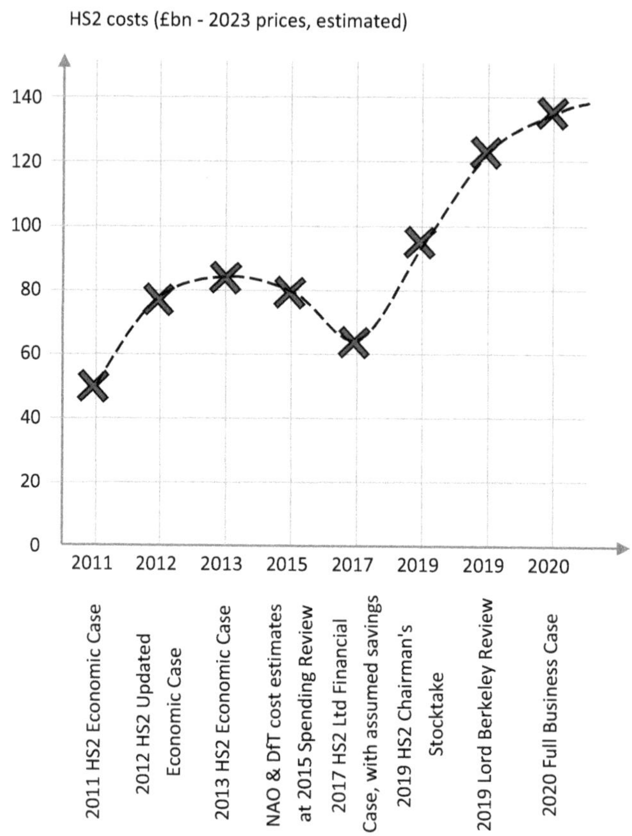

Figure 7.4 HS2 increasing cost estimates in 2023 prices.[12]

80 pence for every £1 spent[11] representing a decrease in benefits of just over 65%. A dramatic reduction in the Cost Benefit ratio would normally be caused by factors such as static benefits in the face of rising costs, changes within the project environment or poor results from the estimation process.

The Institute for Government (IoG)[12] cite four reasons for the increase in costs. They say that scope changes account for £1bn of the £20–26bn increase between 2017 and 2019. Poorer than expected ground conditions also resulted in additional structural works being required before the track could be laid, and this led to a further £5bn of cost increases, shown in Figure 7.4, between the years 2017 and 2019. Optimism bias (discussed in Section 7.3.5,) and the impact of inflation were the third and fourth reasons cited by the IoG for the cost increase. The press went further, suggesting that fraud may even been a further cause behind the escalating costs.[13] What can be reviewed here though, is the impact of poor estimating on the four reasons identifies by the IoG – scope increases, unexpected ground conditions, optimism bias and inflation.

7.3.3 *How Could a Developed Estimation Process Have Helped?*

One of the scope increase cited by the IoG was the increase in length of the tunnel through the Chilterns, as part of HS2 Phase 1, due to perceived political pressure from protestors and constituents. If the project estimate is to retain its validity, then a material changes such as this, must go through a formal change control process (see Chapter 12) supported by estimates supplied by industry experts. This would have led to the cost and benefit delivered by the proposed change being assessed and if approved, the project budget would have been re-baselined and thereby the project estimate, although increased, would have been approved at the new level. The approval of any new estimate should only be agreed if the additional benefits delivered by the change can be shown to outweigh the additional costs - a mini cost benefit analysis (see Chapter 2). When a cost benefit analysis is not performed, the estimates will often appear to rise uncontrollably and the revised project can lose sight of any newly proposed benefits, and hence the reasons behind the increased scope.

The cost growth driven by the unforeseen ground conditions noted by the IoG[16] could have been avoided had a more thorough method of estimating been used at the start of the project. It is not clear which estimating method was used, whether it was subjective, comparative, parametric or analytical. Whichever method was used, it appears to have been ignorant of the ground conditions suggesting that a detailed bottom-up process was not used. The factors relating to the ground conditions could have been identified and used to provide a more detailed and accurate estimate, which given the increased financial cost involved, would seem to have been the correct course of action. Of course, to undertake the work needed for a full ground survey across the complete route of HS2 would have taken time and would have been expensive, and could, theoretically, have produced the same estimate as the estimate that was produced. The benefit of an early 'correct' estimate is that the initial invest/don't invest decision could have been made before any sunk costs were incurred and hence the costs and time incurred in cancelling the project could have been avoided. If this had been the case, then the additional sums spent on providing estimates would have represented value for money.

The third and fourth of reasons for the increase in project scope cited by the IoG were optimism bias and the impact of inflation. With any long-term project, it can be assumed that prices or costs of delivery will be time sensitive and increase as the years pass. If a project costs are to be incurred, for example, in 10 years' time, then an inflation factor should be factored in. The IoG report that 'the rise in prices and labour costs since the early estimates has had a 'significant' effect on planned costs'.[12] If an inflation factor had been included in the project estimates, then it would appear that the factor was too low and that it should have been greater. In other words, it was too optimistic. The IoG chronicles that 'the Government was too optimistic about how cheaply and quickly it could build HS2.'[12] This is the problem of optimism bias evidenced. Any estimate that a project sponsor or project manager receives should be validated and tested. It is testing and validation of estimates provided by, for example, project contractors than can counteract any undue optimism bias, or any low estimate that might be put forward in order to get a project approved or to win a contract. It is important to also consider the build time for HS2. The IoG say this too was underestimated, which means that the delivery of the project benefits was accelerated in the business case but would actually have been delivered later than forecast. If benefits are late in delivery, their value is directly impacted as they will either have a shorter in use period or require discounting at a higher rate, or perhaps both. This will directly impact any business case.

Project estimates received must therefore be tested, as far as possible, for reasonableness and completeness. An issue can arise where a project initially starts without the full knowledge of its constituent parts, as in the case of HS2. This issue can be approached through the use of a three-point estimating approach, and in fairness to the HS2 project this type of approach was followed in the 2020 full business case,[10] where the DoT talk of the cost estimate range, or funding envelope, for Phase One being £35–45bn and proposed a target cost of £40 billion, essentially a minimum, maximum and most likely cost. This estimate drove the total project forecast cost up to £108.9bn (in 2015 prices). The reason quoted by the IoG for the range estimate was the 'uncertainty over such a complex programme (group of related projects)'. Uncertainty will always be present when estimates are made, but it is the complexity of projects that means the most appropriate and considered approach to estimating must be taken to reduce the impact of this uncertainty where ever possible and to help in reducing project failures.

7.3.4 Conclusions on HS2

The Infrastructure and Projects Authority (IPA) 2022–2023 Annual Project Report[14] gave both HS2 Phase 1 and HS2 Phase 2a a 'red' rating. The IPA's definition of a 'red' rated project is that 'successful delivery of the project appears to be unachievable. There are major issues with project definition, schedule, budget, quality and/or benefits delivery, which at this stage do not appear to be manageable or resolvable.' Whilst this is a general description of a 'red' project the factors relating to budget and schedule not being delivered as estimated are clearly evident. These could have been avoided if Freidman's guidance was followed.[17] Freidman suggested that under estimates lead to disaster, over estimates become self-fulfilling whilst realistic estimates minimise final project costs. Accurate estimates therefore are the only really useful estimates and so time and thought needs to be put into their formulation.

With respect to HS2, the IoG concludes that 'Government needs to make sure the Treasury has experts in place who can scrutinise cost estimates from departments and tackle the optimism bias that continues to creep in when costing construction projects. Policy makers should also take a more flexible approach to planning. Rather than trying to plan for every detail over a long timescale, ministers and officials should consider starting with a minimum viable option and then adding to it over time to avoid locking themselves into an option that ends up being unexpectedly and unnecessarily expensive.'[15]

7.4 Project Estimates Before the Birth of Modern Project Management*

(* In this section, the research relating to the construction of HMS Barham 1912–1915 mentioned in the introduction is drawn upon to help illustrate how the subject of this chapter has developed.)

Whilst no records of estimating process were found within the HMS Barham archive papers, a fascinating insight into the process was found in the papers of John Brown and Co. who built HMS Barham, relating to the battlecruiser HMS Hood (1918) and an unnamed new 'proposal' vessel.

The extract from the Glasgow University Scottish Business Records shows both a comparative and analytical approach to estimating.

On the left-hand side of Photo 7.1, high level measurements are provided and compared between the two vessels. On the right-hand side, estimates for the cost of a variety of different types of wood are given showing the cost figures for both vessels with an uplift in the 'proposal' figures reflecting the increased size and 'rate'; or unit price.

This type of approach to estimating, comparative and analytical, can therefore be seen in use before the formulation of Modern Project Management. If a larger data base were to have

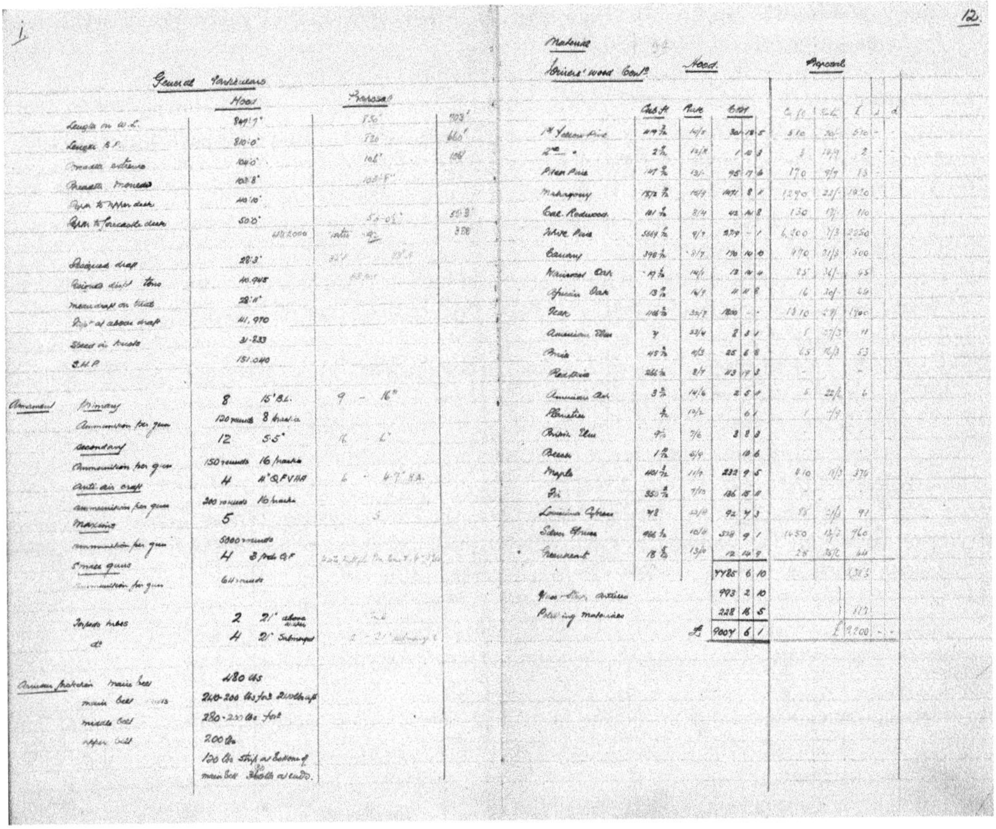

Photo 7.1 Estimation extract from Scottish Business records circa 1918. By permission of the National Records of Scotland and University of Glasgow Archives & Special Collections, Records of Upper Clyde Shipbuilders Ltd, Clydebank Division collection, GB248 UCS 1/85/3.

existed at the shipyards, then there would have been the potential for the constructors to provide parametric estimates, where a formula could have been developed to cost each type of wood required dependent on, for example the length and breadth of each new vessel etc.

References

1 Project Management Institute. (2021). *The standard for project management and a guide to the project management body of knowledge* (7th ed.). Project Management Institute, Inc.
2 APM. (2019). *APM body of knowledge* (7th ed.). Association for Project Management.
3 Ohara, S. (2017). P2M – A guidebook of project & program management for enterprise innovation (3rd ed.). Project Management Association of Japan (PMAJ).
4 HS2. (2019, August). HS2 chairman's stocktake. https://assets.publishing.service.gov.uk/media/5de7d73b40f0b60881a07230/hs2-chairmans-stocktake.pdf [Accessed 12 June 2024].
5 House of Commons Library. (2019). High Speed 2: The business case, costs and spending. https://commonslibrary.parliament.uk/research-briefings/cbp-8601/ [Accessed 6 June 2024].

6 Greening, J. (2012, January 10). High speed rail. https://www.gov.uk/government/speeches/high-speed-rail--30 [Accessed 17 June 2024].
7 HS2. (2024). Trains. https://www.hs2.org.uk/building-hs2/trains/ [Accessed 4 June 2024].
8 Department for Transport. (2017). High speed two from concept to reality – Moving Britain Ahead. https://assets.publishing.service.gov.uk/media/5a81d343e5274a2e87dbfae9/high-speed-two-from-concept-to-reality.pdf [Accessed 14 June 2024].
9 Department for Transport. (2011). High speed rail consultation summary of the HS2 economic case. https://hs2ltd.wordpress.com/wp-content/uploads/2011/05/summary-economic-case_0.pdf [Accessed 15 June 2024].
10 Department for Transport. (2011). Full business case high speed 2 phase one. https://assets.publishing.service.gov.uk/media/5fc51d7b8fa8f54755392863/full-business-case-hs2-phase-one.pdf [Accessed 15 June 2024].
11 Institute for Government. (2023). HS2: Costs and controversies. https://www.instituteforgovernment.org.uk/explainer/hs2-costs [Accessed 16 June 2024].
12 Department for Transport. (2012). High speed rail: Investing in Britain's future – Decisions and next steps. https://assets.publishing.service.gov.uk/media/5a789f5540f0b63247698da2/hs2-decisions-and-next-steps.pdf [Accessed 15 June 2024].
13 Calvert, J., & Arbuthnott, G. (2023, October 21). The secret files that expose a multibillion-pound cover-up at HS2. *The Sunday Times*. https://www.thetimes.com/uk/politics/article/hs2-billion-pound-coverup-cost-files-investigation-skzv2nxwj [Accessed 18 June 2024].
14 The Infrastructure and Projects Authority. (2023). The 2022–23 annual report on the government major projects portfolio (GMPP) from the infrastructure and projects authority (IPA). https://assets.publishing.service.gov.uk/media/64c91eaed8b1a71e86b05df3/IPA-Annual-report-2022-2023.pdf.pdf [Accessed 14 June 2024].
15 Institute for Government. (2023). The latest HS2 delays expose deeper problems with UK infrastructure planning. https://www.instituteforgovernment.org.uk/comment/hs2-delays-infrastructure-planning [Accessed 17 June 2024].
16 Institute for Government. (2021). HS2: Lessons for future infrastructure projects. https://www.instituteforgovernment.org.uk/publication/hs2-lessons-future-infrastructure-projects.
17 Freiman, F.R. (1983). The fast cost estimating models. American Association of Cost Engineers. *AACE Transactions, Philadelphia.*

Chapter 8

Scheduling

8.1 The Purpose of Scheduling

The APM define a schedule as a timetable showing the forecast start and end dates for the tasks within a project.[1] Much is implicit within this definition once it is acknowledged that there are connections and constraints between project tasks. The PMI[2] make this explicit stating that 'a schedule is a model for executing the project's activities, including durations, dependencies, and other planning information.'

P2M[3] illustrates the connection between establishing a lifecycle (see Chapter 4) and the schedule itself, stating that in order to create a project schedule, it is first necessary to know the project's beginning and end dates and the phases contained within. The purpose of the schedule is to describe the basis of project delivery and it will draw on the knowledge of those involved in the project. Once the project schedule starts to be delivered it will be impacted, either positively or negatively, by all the project management processes discussed in this book. For instance, the way that risks are managed (Chapter 9) will impact the schedule, the way project communication is delivered within the project (Chapter 6) and the way in which the project is controlled and monitored (Chapter 11) will all impact the project schedule. Once formulated, the project schedule along with the project budget (Chapter 10), becomes a key project baseline or document for tracking progress.

Similar to the views expressed by P2M, the APM[1] say that scheduling is a collection of tools and techniques used to show when the work needed to deliver a project is required to take place, and that in order to do this the relationship and dependencies between project tasks need to be understood. It is these tools and techniques that are analysed in Section 8.2 below.

8.2 How Is the Project Schedule Delivered, Managed and Used?

Firstly, the project plan is not the project schedule. The project schedule is part of the project plan, but the plan will also include other documents that describe how the project will be run and delivered, such as the Risk Management Plan, the Communication Management Plan, the business case etc. The purpose of the project plan is to add as much detail and certainty to the project as possible at this stage and then to communicate the project aims, objectives and benefits to the project team, stakeholders and wider if required.

When the project schedule is discussed, people often expect to see a document similar to the simple construction project Gantt Chart shown below (Figure 8.1):

DOI: 10.1201/9781003470748-9

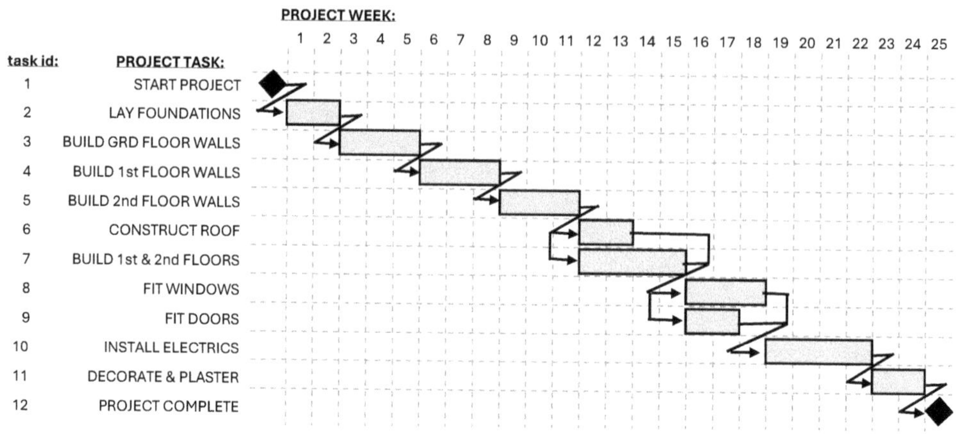

Figure 8.1 A simple Gantt Chart.

The Gantt Chart lists the project tasks on the left-hand side and represents their delivery position and duration across the project's lifecycle by horizontal bars aligned with the project timeline at the top of the diagram. The black diamonds represent milestones, or zero duration events that mark a key stage in the project's delivery, additional milestones could be added during the project delivery rather than just at the beginning and end of the project. The last thing to note in the diagram, is the arrows or linkages between the tasks showing which needs to be completed before the next can begin.

The question though must be, how is this schedule or Gantt Chart developed? What is the process behind its construction?

8.2.1 *How Is a Project Schedule Constructed?*

If a project was tasked with restoring a 1950s sports car and putting it back on the road, what would it look like? How could a project schedule be established? Firstly, the project would need to be discussed with the project funder – the project sponsor or client.

The conversation would centre around what was required to move from the left-hand side of Photo 8.1 below, to the right-hand side.

There are four key stages to developing a project schedule and each has a definitive output. These are, in order, the Project Breakdown Structure, the precedence table, the Network Diagram and the Gantt Chart itself.

8.2.1.1 The Project Breakdown Structure

The project needs to be broken down into sub-parts or tasks so that these tasks can be allocated to teams or individuals and so that the progress of the project can be measured against them.

There are two ways of doing this, either by describing the outputs of a project, or by detailing the jobs, tasks or work-packages that need to be completed in order to deliver the project. By far the most common of the two, and the most descriptive in terms of enabling the allocation of the work-packages, is the second of these, commonly termed the Work Breakdown Structure or WBS.

The WBS includes all elements of work which must be completed in order to deliver the agreed project benefits. It therefore describes the scope of the project. A defined and agreed

Photo 8.1 The restoration project.

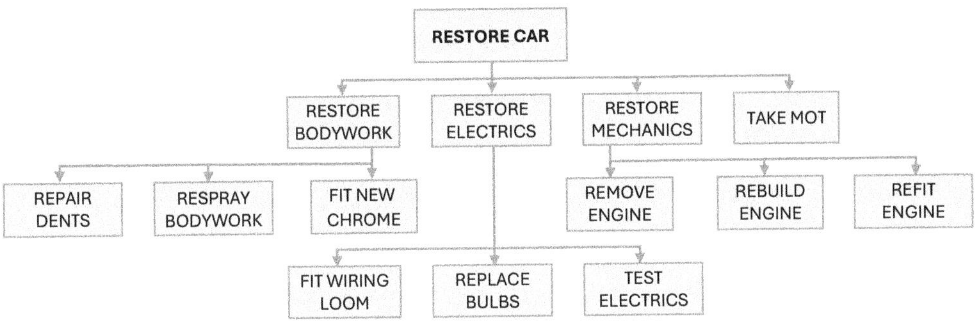

Figure 8.2 A simple WBS.

scope definition is a key enabler of successful project delivery. Without it the project deliverables and purpose will be unclear, relationships between different parts of the project may be uncertain and there will not be a firm contract agreed between the project sponsor and the project team thereby increasing the chance of scope creep and making the monitoring of progress against a baseline more difficult.

A simplified WBS covering a car restoration project is shown in Figure 8.2.

It can be seen from Figure 8.2 that the WBS has a hierarchical structure. The first level is the project itself which breaks down into four level 2 tasks – restore the bodywork, electrics and mechanics of the car and then pass the MOT test. The level 2 tasks though are too general for a team to start work on them. For example, what does 'restore the bodywork' actually mean? How much needs to be done and when does the project stop this work? To answer this the work needs to be broken down into more granular elements, so at level 3 the task of 'restore bodywork' is defined as repair the dents, respray and fit new chrome. The breakdown can be continued through further levels as required. The aim is to break the project down into tasks small enough and defined enough to enable a small team to be able to deliver them and to enable the project manager to monitor and control the project against in terms of time, cost and outputs.

The WBS should be read top down rather than from left to right – there is no inference of dependency or timings between the tasks it is simply a vertical breakdown of the tasks from top to bottom. The only tasks that need to be completed are those that are not broken down further. Looking at the WBS at Figure 8.2, there are ten tasks which are not broken down further (including 'take MOT'). If these and only these are completed then the project, as described, will have been delivered.

8.2.1.2 The Project Precedence Table

Once the scope of the project has been defined, the relationship between the tasks involved needs to be studied. Some of this analysis may be intuitive, some of it may require discussion with project team members. Either way, the final result should be agreed with those responsible for completing the tasks. The precedence table is a way of recording the sequencing of the project tasks in order to successfully deliver the project in a timely and cost-effective manner.

Figure 8.3 shows the relationship between the tasks. Task 1 has no preceding task and so can be started straight away. Task 2, the engine rebuild can only be begun once Task 1 is complete, hence '1' is shown as the preceding task in the third column. Refitting the engine has two tasks that must be completed before it can be started. The engine, obviously, must be rebuilt (Task 2,) and the body must be ready to receive it, with Task 6 completed. The remaining tasks follow the logic of the precedence table in a similar fashion, until the project is complete.

The next project artefact that needs to be addressed is the project network.

8.2.1.3 The Project Network

The project network is derived directly from the precedence table and no additional information is required to complete it. The network shows in a diagrammatic format how the tasks are linked together and provides exactly the same information as is displayed in Figure 8.3. There are three ways which a task could be connected to another. By far the most common is where a task needs to finish before the next task begins. This is known as a Finish to Start Relationship and this is what is shown in the column headed 'Preceding Task(s)' in Figure 8.3. Occasionally relationships such as 'Start to Start' or 'Finish to Finish' may be present between two project tasks but these are seen far less frequently than the 'Finish to Start' relationship.

Figure 8.4 shows the project network which includes the same ten tasks as detailed in the precedence table plus two markers, or 'milestones', the project start and finish, shown in grey. Once the network is completed it can be used to calculate both the best case project duration and also the critical path, or sequence of tasks which determine that best case duration. This is called network analysis.

TASK No.	TASK DESCRIPTION	PRECEDING TASK(s)
1	REMOVE ENGINE	-
2	REBUILD ENGINE	1
3	REFIT ENGINE	2,6
4	REPAIR DENTS	1
5	RESPRAY BODYWORK	4
6	FIT NEW CHROME	5
7	FIT WIRING LOOM	6
8	TEST ELECTRICS	7
9	REPLACE BULBS	8
10	TAKE MOT	3,9

Figure 8.3 A precedence table.

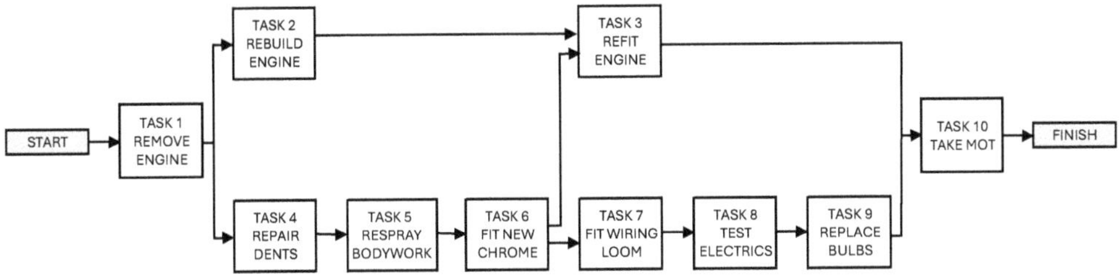

Figure 8.4 The project network.

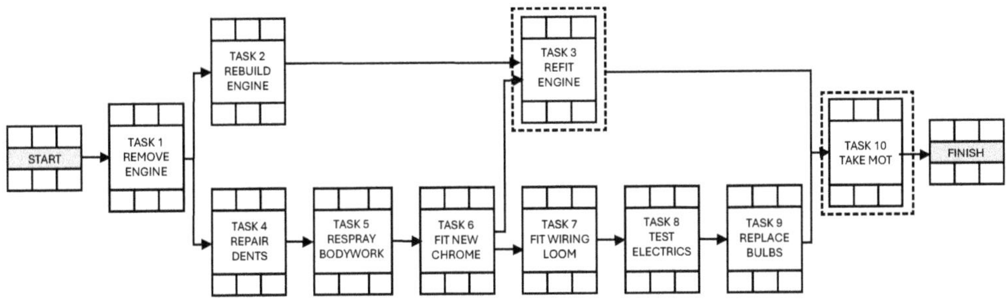

Figure 8.5 The project network augmented.

The first stage in doing this is to take the network in Figure 8.4 and simply add another three boxes above and below each task box, as shown in Figure 8.5.

It is also noticeable that in Figure 8.5 the 'Refit Engine' and 'Take MOT' tasks have a dotted line around them. This is to draw the reader's attention to the particular nature of these tasks. They are the only tasks in the network which have more than one predecessor task feeding directly into them. This will become important as the network analysis is completed and such points within the network are known as 'points of convergence.'

The purpose of the three boxes both above and below each task that have been added in Figure 8.5 is to allow specific task information to be calculated and added to the network. The precise nature of this information is shown below in Figure 8.6.

The earliest a task can start, its duration and the earliest a task can finish must be noted for each task. These can be seen across the top row of Figure 8.6. Across the bottom row, the latest a task can start, its possible float and its latest finish are entered. The one term here that needs explanation is 'float'. Within a project network float describes the amount of time that the start of a task can be delayed without impacting on the end date of the project. This will become clearer in the discussion below.

In order to complete the network analysis, three separate phases must be completed. These are:

1 The Forward Pass. This calculates both the earliest start and finish.
2 The Backward Pass. This calculates both the latest start and finish.
3 Critical path and float calculation.

The phases must be calculated in the sequence shown, phase 1 through to phase 3.

In order to do this, the first thing needed is a list of durational estimates for the project tasks. These estimates are derived using the approaches described in Chapter 7 and are often provided

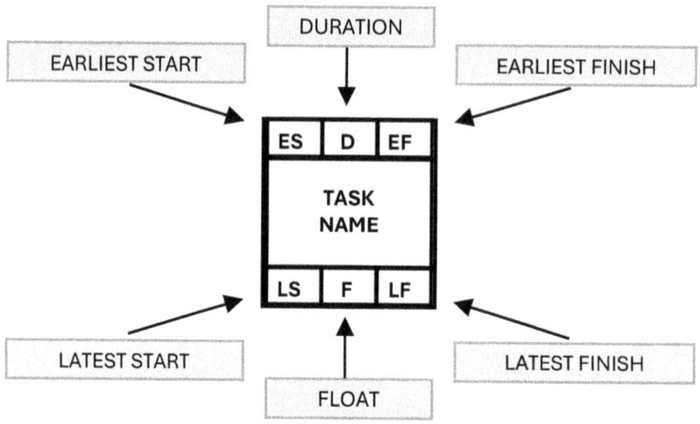

Figure 8.6 The content of each of the boxes added to the project network.

TASK No.	TASK DESCRIPTION	PRECEDING TASK(s)	TASK DURATION (DAYS)
1	REMOVE ENGINE	–	2
2	REBUILD ENGINE	1	15
3	REFIT ENGINE	2,6	3
4	REPAIR DENTS	1	5
5	RESPRAY BODYWORK	4	5
6	FIT NEW CHROME	5	1
7	FIT WIRING LOOM	6	3
8	TEST ELECTRICS	7	1
9	REPLACE BULBS	8	1
10	TAKE MOT	3,9	1

Figure 8.7 The precedence table with durational estimates added.

by the team members responsible for that stream of tasks. They can be agreed and recorded in an extra column added to the precedence table. (See for example Figure 8.7.)

Once the estimates are agreed, they can be entered into the top middle box of each task in the network diagram (Figure 8.8).

The duration of the start and finish milestones is zero as these are not tasks but simply progress markers within the network.

The network is now ready for the Forward Pass to be undertaken.

The very soonest that the project can be begun is now, at time zero, so the one other figure that is known at this stage, is the earliest start for the start milestone. Once this is inserted in the network, the earliest finish can be calculated by adding the duration to the earliest start. In this case the sum is straight forward: 0 + 0 = 0. If the start milestone can be finished at time zero, then at this very same moment in time, the next task in the network ‘remove engine’ can be begun, this takes 2 days so if the earliest it can start is 0, then the earliest it can finish is the end of day 2 (0 + 2 = 2). Once the engine is separated from the chassis, the end of day 2 then becomes the

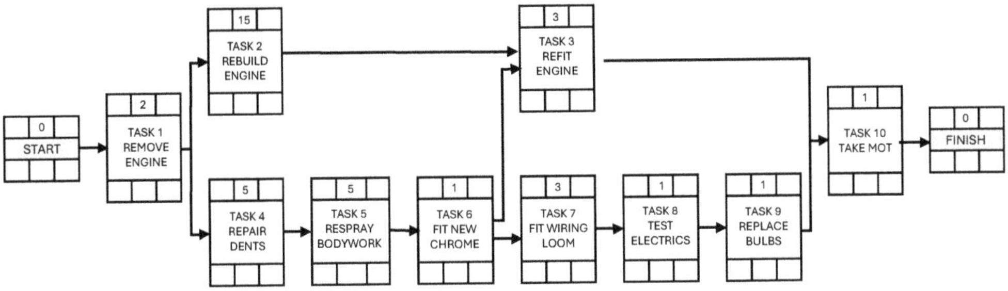

Figure 8.8 The project network with durational estimates added.

earliest that the two subsequent tasks, 'rebuild engine' and 'repair dents' can start. The earliest finish for these two tasks can then be calculated, using the equation:

$$Earliest\ Start + Duration = Earliest\ Finish$$

For task 2 and task 4 this equals 2 + 15 = 17 and 2 + 5 = 7 days respectively. It can be seen that the earliest finish for a task becomes the earliest start for the next task until a point of convergence is reached.

What figure should be inserted as the earliest start for Task 3, Refit Engine?

According to the network, both 'Rebuild Engine' and 'Fit New Chrome' must be completed before Task 3 can start. The tasks are not both due to be completed until day 17, therefore Task 3 cannot begin until day 17 at the earliest. At a point of convergence on the Forward Pass the rule is that the highest preceding earliest finish becomes the earliest start for the next task. Using this rule, the Forward Pass can be completed to the first point of convergence as shown in Figure 8.9, noting the second point of convergence following tasks 3 and 9 as they link to 10.

Having completed the Forward Pass, it gives the project manager one piece of vital information – when the project can be completed by. Figure 8.10 shows that, if the work breakdown structure and estimates are accurate, then it can be expected that the project will be completed in twenty-one days.

The Backward Pass, the second step listed above, is the next step needed in order to understand more about the project and its network.

The Backward Pass is begun, from the right-hand side of the network diagram, from the finish milestone. The finish milestone, from the Forward Pass has an earliest finish of 21 days. It is important that if the project can be finished in twenty-one days, then this should be the aim, since any other approach would add time and inevitably cost to the project. To reflect this, the earliest finish of the finish milestone must equal the latest finish for that milestone. In Figure 8.11 the figure of twenty-one days can be seen moved down from the top right of the finish milestone to the bottom right of the milestone.

In order to complete the Backward Pass, the latest start for each task can be calculated using the equation:

$$Latest\ Finish - Duration = Latest\ Start$$

This holds true until a point of convergence is reached such as that highlighted in Figure 8.11 where the Backward Pass focuses on Task 1. Here the latest finish for Task 1 must enable the

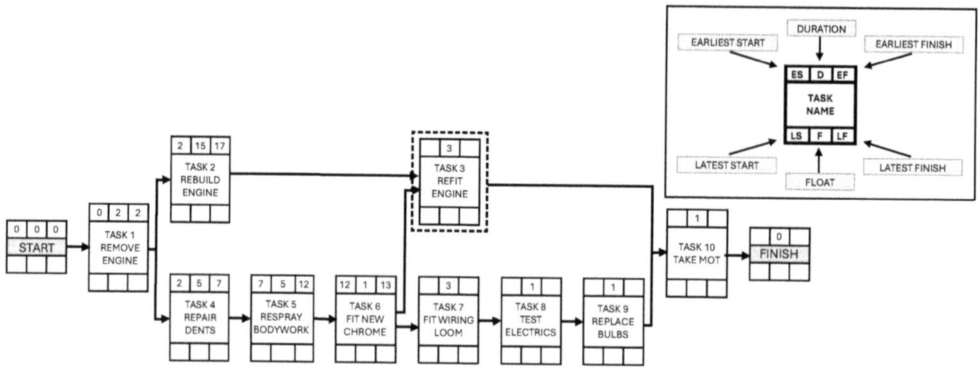

Figure 8.9 The Forward Pass completed up to a point of convergence.

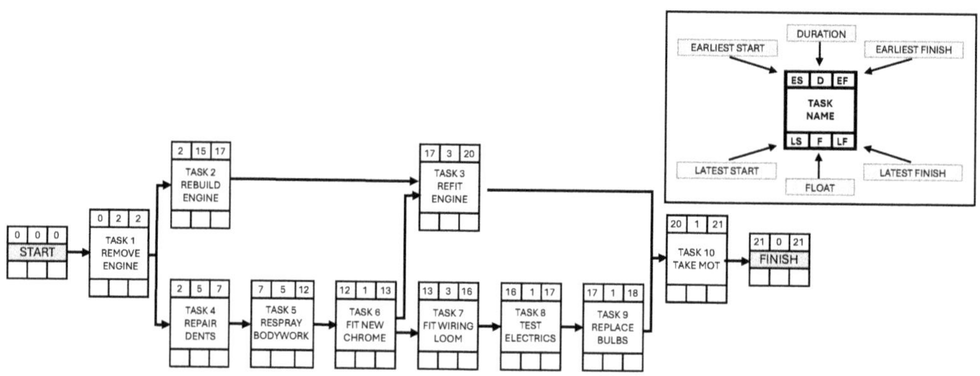

Figure 8.10 The completed Forward Pass.

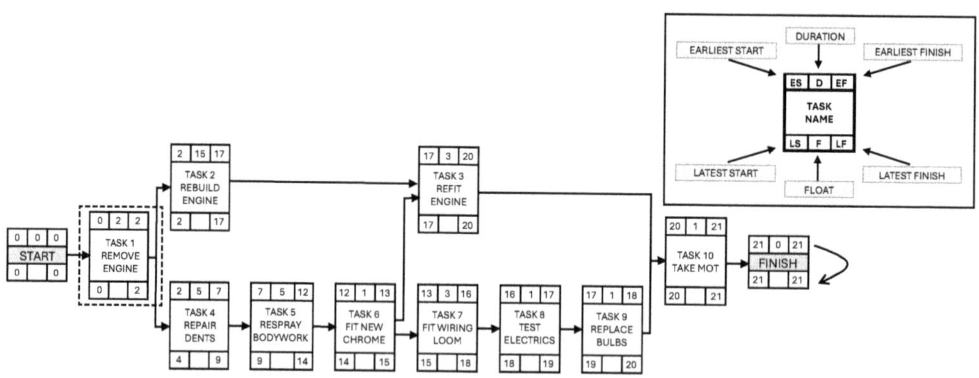

Figure 8.11 The completed Backward Pass, with a point of convergence highlighted.

latest start for both Tasks 2 and 4. These latest starts are shown in Figure 8.11 as days 2 and 4 respectively. If both tasks are to be started on time, then the latest finish for Task 1 must be the lower of these two numbers, consequently day '2' is inserted as the latest finish for Task 1. There is a second point of convergence on the Backward Pass as tasks 3 and 7 link back to task 6.

The Backward Pass in itself, does not tell the project manager too much, that is until the float is calculated. Having completed the Backward Pass, the next step in analysing the project network is to calculate the float, which in terms of the network analysis can be expressed as the amount of time that an individual task can be delayed without impacting on the end date of the project. This is calculated by looking at difference between the earliest start and the latest start. The equation required is:

$$Float = Latest\ Start - Earliest\ Start$$

This is shown in the Figure 8.12 below.

The float in Figure 8.12 is shown in the middle of the bottom row of figures. Tasks 4,5,6,7,8 and 9 all have two days of float available. If one of these tasks (for example Task 4) uses all of the two days of available float, then this will impact the time taken for the task, the duration will increase by two days and so the earliest finish will be pushed back by two days and will now equal the task 4's latest finish. If the network is recalculated to reflect this change, then there will be no float available for any of the subsequent tasks.

Figure 8.12 shows the network with task 4 taking just the estimated 5 days duration, and it can be seen that there is a row of highlighted tasks where the float is zero, (Start–1–2–3– 10–Finish). This row of tasks showing zero float is referred to as the 'critical path' through the project. It means that if any of the tasks on this critical path are delayed or take longer than the duration estimated, the project will not finish on time. As a result, these are the tasks that will need a particular focus from the project manager when it comes to facilitating the tasks and monitoring and controlling them. The network, if correctly constructed, will always have at least one critical path through it, and may on occasion have more than one. If a network is calculated and after the Backward Pass, all the numbers in the start milestone are shown as zero, then this is a good, although not foolproof, indication that the network calculations are correct. If the start

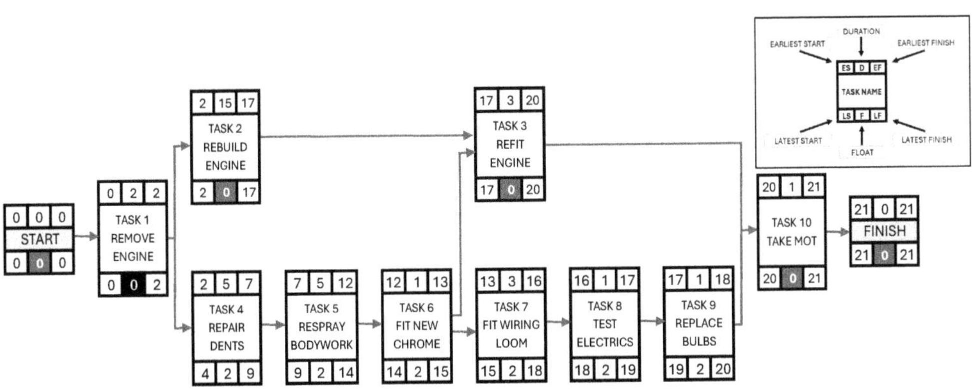

Figure 8.12 The project network showing the float calculation with the critical path, or path of zero float highlighted.

milestone has a figure other than zero in any of the six boxes, then the network calculation is certainly wrong.

8.2.2 Other Types of Project Float and Project Milestones

The float calculated in Section 8.2.1 is termed 'Total Float' and can be defined as 'the amount of time by which an activity in a project network can be delayed without impacting the estimated project duration'.

There is a second type of float which can be available to the project manager, known as 'Free Float'. This can be defined as 'the amount of time by which an activity in a project network can be delayed without impacting the start time of any subsequent task'.

The two definitions might seem fairly similar, but can be useful in quite different ways. Figure 8.13 shows the difference diagrammatically.

In Figure 8.13 Free Float only appears at the point of convergence between Task D and Task E. Free Float will only occur at a point of convergence and in this case, it is due to the start of Task E being constrained by the finish time of Task C. Task C has an earliest finish of Day 8 and Task D of Day 4. Task E must wait until both C & D are complete, so cannot start until day 8. But Task D can finish by day 4. This difference, 8 – 4 is the Free Float. It is free in the sense that it can only be used at a point of convergence, and so there is no cost to other parts of the schedule. If it is not used here, then it is lost. Because of its 'free' nature, it can be useful to the project manager. If whilst monitoring the project delivery Task C appears to be taking longer to be delivered than expected then, assuming the resources are transferable between the tasks, a proportion of resource can be moved to help speed up Task C and to bring it back on target to finish by its latest finish date of day 8. This will of course impact the speed of delivery for Task D. Task D has

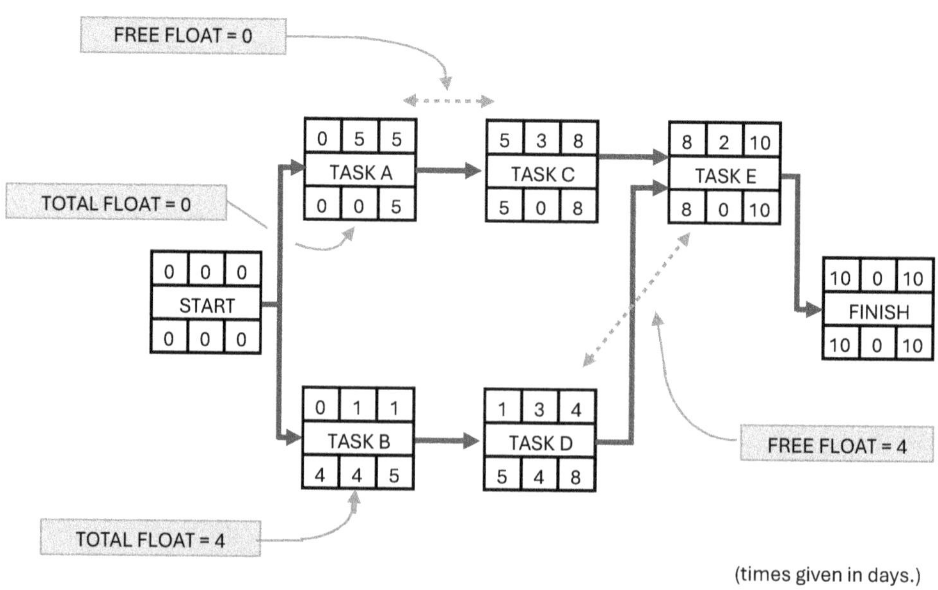

Figure 8.13 Total float and Free Float within a project network.

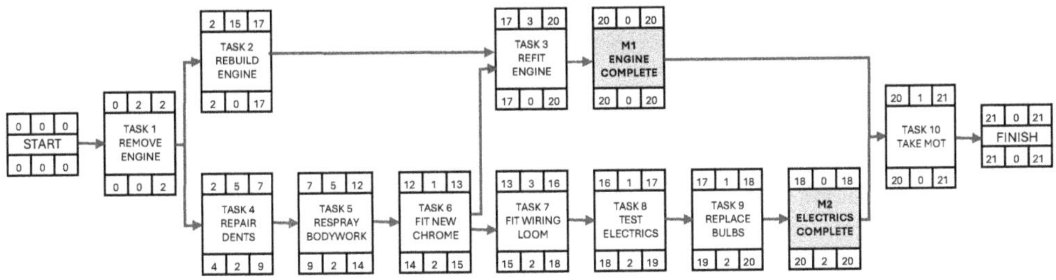

Figure 8.14 The project network with milestones added.

a duration of 3 days. If half of its resource was transferred to Task C, then all things being equal, this would mean that Task D would take 6 days and finish on day 7. Both tasks would now be complete by day 8 in order for Task E to begin on time without consuming any of the remaining Total Float in the network.

In order for the project manager to be able to identify these opportunities at the earliest stage the project network must be monitored. This can be helped by the insertion of 'milestones'. The start and finish milestones have been discussed previously and defined as having a duration of 'zero' and so no impact on the project's overall duration. As milestones have no impact on overall project duration, further milestones can be added wherever they are useful without impacting on the projects end date. A milestone can be useful to both the project manager and the client to prove progression of the project according to plan or to perhaps highlight some deviation from the schedule. Milestones can be added at any point in the network, but this is best done where something tangible is produced as evidence that the milestone has been achieved.

Figure 8.14 shows the project network with the addition of two milestones. In one sense the two milestones, shaded in Figure 8.14 are well chosen. They are very tangible things and hence easy to verify. For instance, the engine can be started and revved, and the headlamps can be switched on. However, in terms of the amount of time left to correct the project should the project fail to meet them, they leave much to be desired. The first of the milestones are not due to be achieved before day 18 of a 21-day project. If milestone tracking is the primary way in which the project is being monitored this leaves very little time for corrective action. Milestones should be spread, roughly evenly, through the project lifecycle so that there is sufficient early warning should anything go wrong with the project.

8.2.3 *How Does the Project Network translate into a Project Gantt Chart?*

A Gantt Chart contains the same information, project tasks, durations and linkages as the analysed project network does.

It is however presented in a different, more familiar format. The Gantt Chart takes the information in Figure 8.14, noting the less than ideal positioning of the milestones, and represents it with a timeline across the top and the tasks listed vertically on the left-hand side (Figure 8.15):

The Gantt Chart can be used to show resources required against each task and then can be used to develop a project budget as shown in Chapter 10.

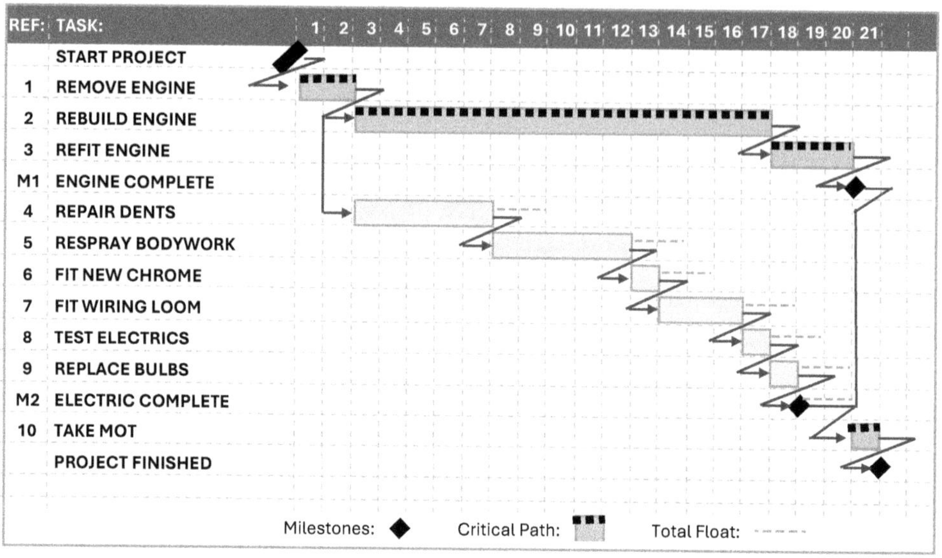

Figure 8.15 The project Gantt Chart.

8.3 Project Review: The Impact of Poor Scheduling on the Project: Queensland Health Payroll Project

8.3.1 The Project Background

It has been said that the 2010 payroll system implementation project at Queensland Health was one of the most spectacular failed technology project failures ever run in the southern hemisphere and it was arguably the second worst failure of public administration in the history of Australia. Queensland Health provides public sector healthcare to the Australian State of Queensland, including the provision of dental, medical, and age-care facilities in the State, the population of which is the most geographically widespread of all the states in Australia,[5] having a population of 5.07 million across an area of 1.85 million square kilometres.

The Payroll Project had its foundations in the 2002 Queensland Government "Shared Services Initiative", the purpose of which was to rationalise government services across a number of different government organisations. The model had been seen to work in large American corporations and had seemingly been applied with success to the public services sector.[6] Overall, the project cost was AU$64.5M in terms of direct cost, and an additional AU$1.2B for 8 years of subsequent operation.

The project was expected to reduce costs and to improve standards within functions such as purchasing, debt recovery, record keeping and paying public servants from numerous departments, covered by the initiative. The aim was to lower the staff running costs for the combined functions and to deploy a set of standard business rules thereby enabling a common set of IT products and programmes and simplifying the systems whilst reducing running costs.

The project was however complicated by the sheer size of the task. Queensland Health employed around 85,000 staff at the time, across a number of different areas and occupations. There were also about 200 different allowances all needing to interface with six different industrial agreements. It was estimated that this resulted in approximately 24,000 different pay combinations.[5]

8.3.2 The Project Issue

The headline issues for the project were twofold. The first being that whilst the contract price for the system itself was AU$6.1m, this increased by a factor of four and the ultimate price was AU$25.7M. The wider programme costs increased to AU$181M which was almost double the price which the main contractor had originally estimated in their tender.

Secondly, the interim replacement solution was originally scheduled to be in place by 31 July 2008. After a series of required changes, the go live date of 4 March 2010 with the first pay day of 24 March 2010 was approved.[6] Go live though wasn't the end of the project issues, the system malfunctioned on delivery and took a number of months until the point was reached where employees were paid accurately and on time. As at 31 May 2012 this required over a thousand payroll employees in order to conduct fortnightly pay runs. This would suggest that the system wasn't operating as originally intended at the go live date, and it has been estimated that going live with the premature system resulted in more than 35,000 payroll mistakes, costing the State up to this point, more than AU$400 million just to operate the system. KPMG estimated that the cost of making the system function for the next five years would be another AU$836m.[4]

The system issue log as of May 2012 had approximately 570 issues logged in it, of which just under 15% had the potential to impact staff pay. Any resulting fixes that were needed were limited in that they could only go live in one of three specified periods each year.

Problems with this system meant that as of May 2012, staff had been overpaid by AU$112.3m,[5] some of which had been repaid whilst some amounts had been waived. Associated with the overpayments, Queensland Health were also required to fund a tax liability which at one stage amounted to an additional AU$1.7m every fortnight.

Not only were costs increasing exponentially, but the hoped for benefits that were expected to be realised through common IT systems, did not materialise either. Pay information and data from across the districts continued to be provided in the formats developed locally in each of the districts. The information was now sent to a centralised team who did not have the benefit of the local payroll officers knowledge, creating additional issues.

The impact of the project failings was felt not just by the individuals served by the new payroll system, who could not afford the financial consequences of a dysfunctional payroll system,[6] but also due to the size of the problems, by the State itself. The Commission of Inquiry Report[6] reflected on the delays, additional costs and impact caused by the system's poor performance, saying that there were 'terrible consequences for the employees of Queensland Health and equally serious financial consequences for the State'. After many months of anguished activity during which employees of Queensland Health endured hardship and uncertainty, 'a functioning payroll system was developed, but it (was) very costly.'

A project such as this, normally has many causes or reasons that contribute to its difficulties, there is seldom just one issue. The Commission of Inquiry Report[6] did however, identify two principal causes of the systems shortcomings. The defects traced back to poor decisions which project officials made when scoping parts of the project and in the subsequent governance of the Project. As a result of these errors the project never delivered the benefits it was supposed to.

The focus of this chapter is of course, on scheduling so the poor project scoping mentioned by the report is of interest. Earlier in this chapter, the scheduling process was reviewed, and the first part of the process discussed was the need to create the precedence table and to identify the project tasks within the project scope. The Queensland Health Payroll System Commission of Inquiry Report[6] established that failings in this area heavily impacted the project. The report stated that 'the scope of the system was inadequately defined and that the scoping activities were expected to be achieved within an impossible timeframe'. The original project scope meant that had the system been delivered as originally scoped, it would have failed entirely to function as a properly defined and constructed payroll system should. The report revealed that the main contractor decided to use the system deployed in the DoH (Department of Housing) as a basis or baseline upon which to construct the new Queensland Health payroll system. The DoH system however did not reflect the complexity of the Queensland Health payroll, Housing and Health were very different agencies in terms of scale, complexity of awards, rostering requirements and software utilised. The report when on to say that the decision to use the DoH solution as a basis for the Queensland Health project was a poor decision driven by a perceived need to complete the project quickly. The nature of the contract between the State and main contractor meant that both parties had to agree the project scope, a process criticised by the Inquiry Report which considered that it would have been preferable for the State to have retained a "casting vote". Without a casting vote, the State was left with the need to agree these matters with IBM who has successful tendered for the contract which was awarded on 5 December 2007.[6] The impact of this was that any agreement from the State related to the project scope was given more by default and neglect than with considered thought.

Inevitably, the scope as it was required changes to be made to it, which resulted in increases in the project price. The lack of a sufficiently developed scope agreed between all parties was a very serious shortcoming. The Inquiry Report laid this out saying 'the system was unsatisfactorily scoped, the State was less than diligent in advising its business requirements to the contractor and negligent in accepting the highly qualified, incomplete and uncertain scope presented to them. In relation to the series of expansions and changes to scope which this led to, it was concluded that most of these ought to have been included from the outset, in the original scoping if the system were to be one which paid staff accurately',[6] which highlights the poor level of scoping and lack of due diligence in allowing the project to proceed unchecked. Part of the difficulty behind the scope definition was caused by the very different understandings which the parties involved had of the initial scope. This confusion, leading to the inevitable cost increase for the project as these views were harmonised, underlines the importance of spending time on producing a workable, and accurate, project definition.

The poor level of initial project scoping was compounded by the fact that changes authorised by the Associate Director-General of the Department of Public Works were agreed without the Associate Director-General fully familiarising themselves with basic documentation relating to the change. Instead, it appears that information from junior officers was relied upon, this lack of engagement resulted in the Inquiry Report stating that it was a miscarriage of responsibility for the Associate Director-General not to have satisfied themselves about the details of the changes they were asked to authorise, particularly as the changes involved the payment of a 'very large sum of public money'.

8.3.3 How Could a Developed Scheduling Process Have Helped?

The findings of the Inquiry Report highlight the lack of certainty regarding the scope involved in the delivery of the Queensland Health project. It can be seen how the undefined scope led to additional project charges as the scope requirements emerged. The late identification of scope requirements meant that increased costs were almost inevitable as some of the work delivered on the project, once the changes were introduced became unnecessary. This meant that not only did work need to be undone, but additional work was required, overall the amount of work done would have been in excess of that required had the initial scope been better defined. This is reflected in the discussion of cost related risk defined in the Inquiry Report which states, when discussing the late development and agreement of scope, that 'despite the problem being a known area of risk, the parties, (the main contractor and the State), were content to proceed notwithstanding they both had very different views as to scope. They thus ignored an area of risk which was known to them and one which they ought to have done more to resolve.'

Had the project scheduling process been based upon a clear and accurate project scope it would have allowed the project to have clear documentation and communication surrounding the business requirements from the beginning of the project.[4] This in turn would have helped provide the degree of certainty required by those delivering the project. Over the approximately five years between the appointment of the main contractor and the system go live date, there was always likely to be change in the project scope, but this should however have been actively managed. In the case of the Queensland Health Payroll Project, the absence of a periodic review of the business needs contributed to subsequent difficulties with system testing and the implementation of a system which did not meet the needs of Queensland Health's operating environment.[4] It was noted by the Inquiry Report[6] that the scope, having been unsatisfactory at the start, remained ill-defined, volatile and unclear, and that there was a lack of management of the scoping process with neither the contractor or client taking any effective measures to define and stabilise scope at any time during the project'. Had the project scoping been thorough from the start of the project and subjected to continuous, rigorous review and change control process, then cost and time would have been saved and the personal hardship suffered by the staff as reported above could have been avoided.

8.3.4 Conclusions on the Queensland Healthcare Payroll Project

The accurate and complete development of the precedence table is vital for any successful project delivery, even more so where the project is as large and complex as the Queensland Payroll Project.

The complexity of the project was no excuse for poor requirement gathering and scope recording. Complexity of project scope means that more time must be put into the definition of project scope and the process of establishing that this definition remains relevant must be maintained. What is clearly wrong within the project is the seeming lack of engagement by the sponsoring body in the change control and scoping of the project, it is hard to believe how anything other than a disaster could have been expected. When considering whether or not the scope is correctly defined reference should be made to the project's business case, and it should be clear how all elements of the benefits required are to be delivered and which project tasks are pertinent to the delivery of each benefit. Benefit Mapping can help significantly in this (see Chapter 12), any tasks without a link to the project benefits should be reviewed and if not contributing to the delivery of a desired benefit removed from the project schedule. In the case of the Queensland Health Payroll Project, it is unclear how the tasks being performed could have linked back to the benefits given how poorly the scope was defined. Given this failing to link tasks with benefits as the first step in planning the scheduling process, the value of any schedule subsequently developed and delivered by the main contractor across the multiple sites and environments involved, must remain questionable.

8.4 Project Scheduling before the Birth of Modern Project Management*

(* In this section, the research relating to the construction of HMS Barham 1912–1915 mentioned in the introduction is drawn upon to help illustrate how the subject of this chapter has developed.)

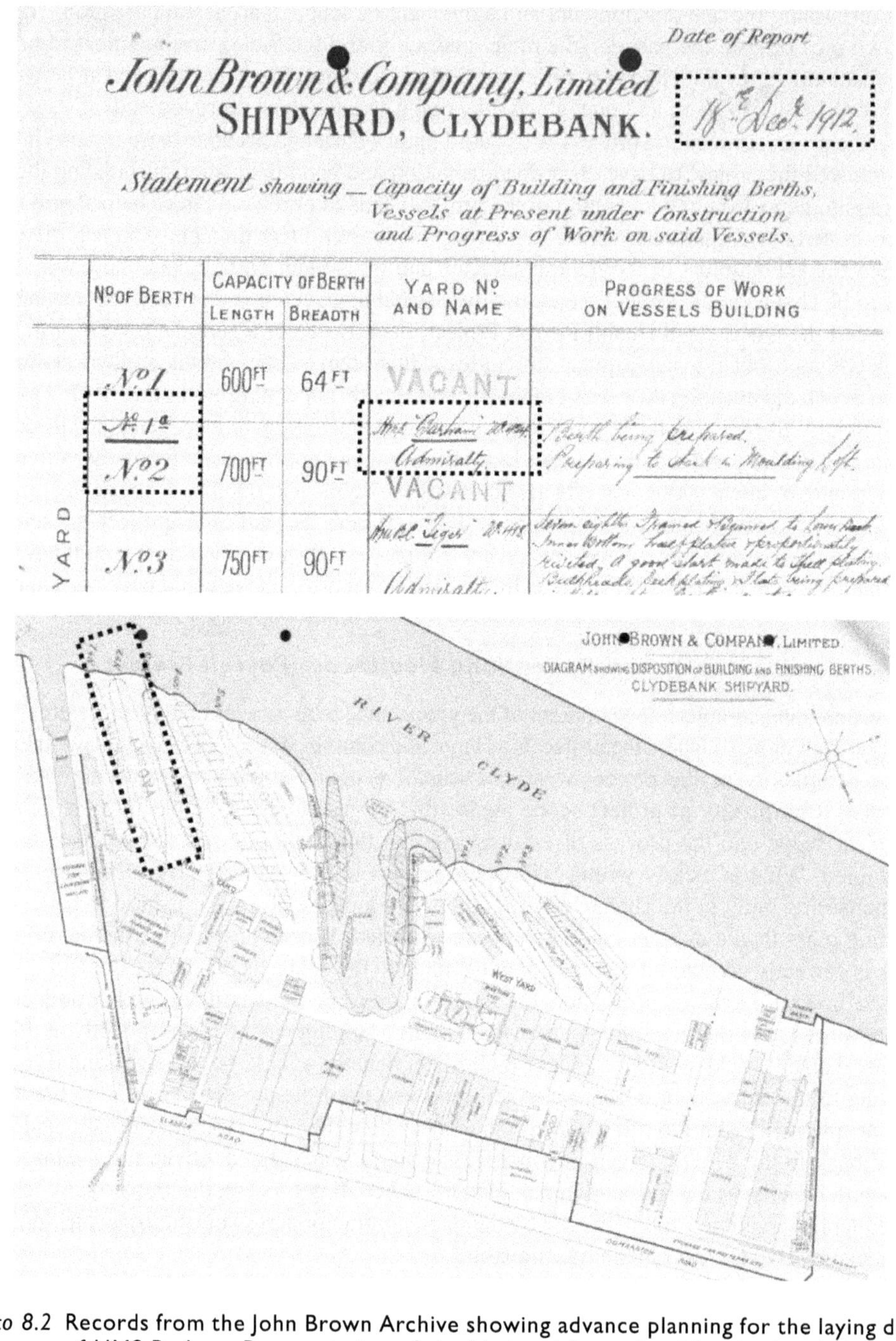

Date of Report

John Brown & Company, Limited
SHIPYARD, CLYDEBANK.

18th Decr 1912.

Statement showing — Capacity of Building and Finishing Berths, Vessels at Present under Construction and Progress of Work on said Vessels.

	No. of Berth	Capacity of Berth		Yard No. and Name	Progress of Work on Vessels Building
		Length	Breadth		
	No. 1	600FT	64FT	VACANT	
	No. 1a / No. 2	700FT	90FT	H.M.S. "Barham" No. [illegible] Admiralty / VACANT	Berth being prepared. Preparing to start in Moulding Loft.
YARD	No. 3	750FT	90FT	H.M.S. "Tiger" No. 418 Admiralty.	Seven eighths framed & [illegible] to Lower Deck. Inner Bottom half plated & proportionately rivetted, a good start made to Shell plating. Bulkheads, Deck plating & Flats being prepared [illegible]

Photo 8.2 Records from the John Brown Archive showing advance planning for the laying down of HMS Barham. By permission of the National Records of Scotland and University of Glasgow Archives & Special Collections, Records of Upper Clyde Shipbuilders Ltd, Clydebank Division collection, GB248, UCS 1/5/11.

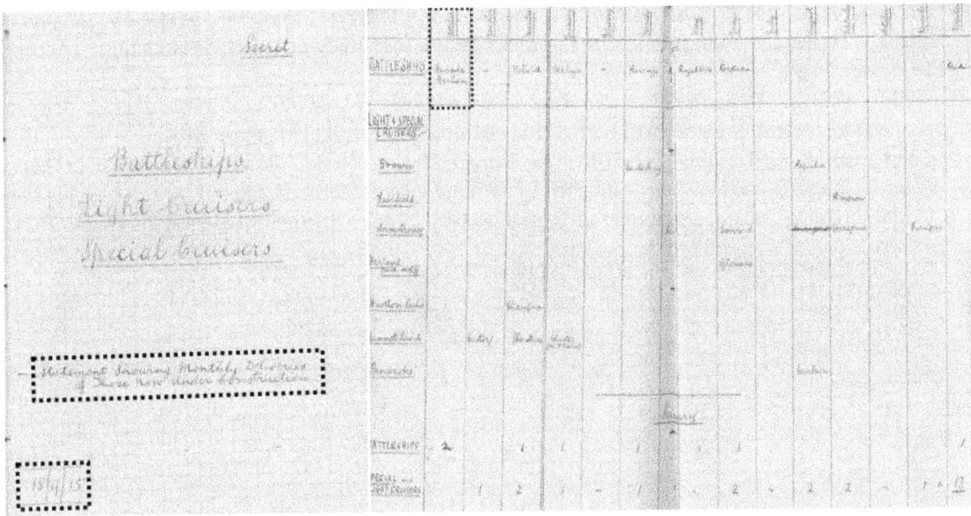

Photo 8.3 Records from the National Archive showing 'deliveries of those (ships) now under construction'. including HMS Barham. By permission of The National Archives ADM 1/8435/297.

Evidence of a schedule presented in the form of a Gantt Chart was not found amongst the HMS Barham project records. This of course does not mean that one was not produced during the 1912–1915 construction of the ship since the records held are incomplete.

Some interesting artefacts were found though, in particularly covering the start and end of the construction period.

The picture at Photo 8.2 shows the berth being prepared for HMS Barham and work being readied in the moulding loft, as of 18/12/1912, (highlighted in the top photograph). This preparation is seen sketched out in the Plan of the shipyard as of the same date, (highlighted in the bottom photograph).

The 18/9/15 date highlighted in the bottom left of Photo 8.3, is prior to the October 1915 delivery date shown for HMS Barham. This shows that a form of advance scheduling or delivery date milestones was produced for the ships shown, covering the period from October 1915 to December 1916.

Evidence is therefore present in the archives that management processes and planning were in place, sufficient to predict future milestone completion dates for projects in progress. This closely reflects the scheduling processes and information requirements that project managers require today.

References

1 APM. (2019). *APM body of knowledge* (79 ed.). Association for Project Management.
2 Project Management Institute. (2021). *The standard for project management and a guide to the project management body of knowledge* (7th ed.). Project Management Institute, Inc.
3 Ohara, S. (2017). P2M – A guidebook of project & program management for enterprise innovation (3rd ed.). Project Management Association of Japan (PMAJ).

4 Henricodolfing. (2019). *Case study 9: The payroll system that cost queensland health AU$1.25 billion*. https://www.henricodolfing.com/2019/12/project-failure-case-study-queensland-health.html [Accessed 19 August 2024].
5 KPMG. (2012). *Queensland health review of the Queensland health payroll system*. https://delimiter.com.au/wp-content/uploads/2012/06/KPMG_audit.pdf [Accessed 19 August 2024].
6 Chesterman R (2013, July 31) 2013 *Queensland Health Payroll System Commission of Inquiry Report*. https://www.healthpayrollinquiry.qld.gov.au/__data/assets/pdf_file/0014/207203/Queensland-Health-Payroll-System-Commission-of-Inquiry-Report-31-July-2013.pdf [Accessed 20 August 2024].

Chapter 9

The Management of Risk

Risk management is a vital part of a project's delivery. Poor risk management will put the agreed baseline time, cost and quality at considerable threat and potentially produce even more serious such as project failure.

9.1 The Purpose of Risk Management

The PMI[1] define a risk as 'an uncertain event or condition that, if it occurs, has a positive or negative effect on one or more project objectives.' Similarly, the APM state that a risk describes the impact that a situation or event can have on the delivery of a project's objectives. The purpose of risk management is therefore to exercise control over these risks so that the overall effect on the project objectives or outcome is optimised. The APM[2] describe risk analysis and management in a little more detail saying that it helps the understanding of both individual risks and the overall level of project risk, and through proactive management helps to optimise project success by minimising threats and maximising opportunities.

P2M[3] makes the point that risk should first be considered at the earliest possible stage of the project's lifecycle. At the business case stage, there is a need to consider whether the expenditure on the project is worthwhile given the project's risk profile, and hence whether the project should be started. P2M[3] identifies two risk types, 'speculative risk' which can have a positive or negative outcome and 'pure risk' which can result only in a negative impact on the project. It says that most effort is focused on managing 'pure risk' and this is true in a majority of projects where risks are considered as negative events and managed accordingly. Positive risks, sometimes referred to as opportunities, can also be managed through the same risk management process or alternatively through techniques such as Value Management.

9.2 Different Sources of Project Risk and Their Impact

The definitions of risk provided by the Project Management bodies have been reviewed above in Section 9.1. A common, working definition often given to describe the size of a risk is simplistically:

$$PROBABILITY \times IMPACT$$

Here 'probability' refers to the likelihood of a risk occurring and 'impact' refers to the effect, or impact, of the risk on the project should it occur. Note that the 'impact' is not described as either positive or negative, so the definition includes both threats or downside risks, and upside risks,

DOI: 10.1201/9781003470748-10

or opportunities. Different projects can have very different sources of risk. Risk can come from simple errors such as those encountered by the Spanish national rail operator, Renfre, who spent approximately £230m on new trains only to find out that they were too big to travel through some of the tunnels on the network[4]. Risk can also of course derive from complicated projects being delivered by complicated bodies in complicated environments. An example of this is the Space Shuttle project discussed later in this chapter.

Often a project manager's first thought, when considering where project risks will be generated, is to look within the project itself. This is of course a perfectly reasonable first step but it is vital to the project's success that the search for risk does not end here. Risks can and will exist within the project itself, but other risks exist external to the project possibly within the organisation delivering the project itself, or potentially a third category of risk may be those risks that originate from outside the organisation delivering the project.

The first area of potential risk is shown in Figure 9.1 as those risks internal to the project. These can be many and varied and will depend hugely on the nature of the project being delivered. They could range from adverse procurement issues such as the late delivery of key components, financial risks derived for instance from inaccurate estimating, poor scheduling or unpredicted fluctuations in prices, interest rates or even foreign exchange rates. Table 9.1 below describes some potential sources and an example of a potential, although not exhaustive, project impact.

The many potential sources of risk for a project can be seen to give rise to a multitude of possible project issues. All of these however, ultimately will impact on either the time, cost or the quality of the project outputs. A project manager will therefore need a keen focus on managing risks so as to ensure that any adverse impact on the time taken to deliver the project, the cost involved, or the quality of project deliverables is minimised. Variations in these three variables, time, cost and quality, often represented by the initials TCQ, will in turn have a direct bearing on the nature and validity of the business case (see Chapter 2). As the quality of a project output is delivered and used, it results in benefits for the commissioning organisation, and if this changes as a result of an identified risk materialising, then the gap between the costs and benefits of the business case will alter too. Similarly changes to cost and time will also influence the project's cost benefit analysis.

Given that the project risk management process is one that is managed by the project manager, then perhaps the project manager is themselves one of the biggest risks on any project! How then, can these risks be properly managed and their negative impacts minimised?

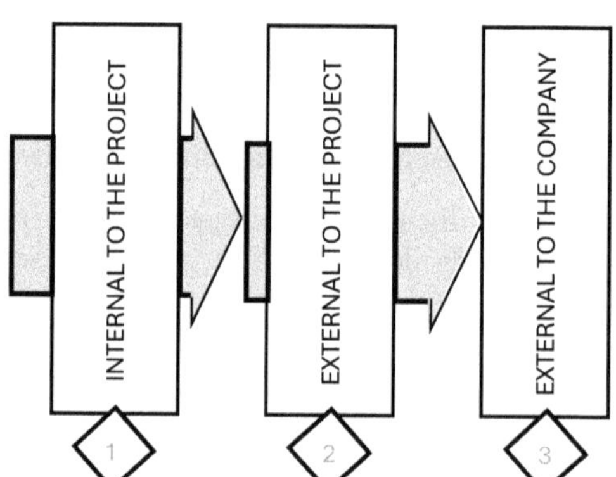

Figure 9.1 – the three potential areas of project risk

Table 9.1 – potential sources and impacts of project

POSSIBLE RISK SOURCE	*POSSIBLE ISSUE*	*POSSIBLE PROJECT IMPACT*
LEGAL RISK	THE PROJECT FAILS TO COMPLY WITH LEGAL REQUIREMENTS SUCH AS HEALTH AND SAFETY LAW.	THE PROJECT COULD BE TEMPORARILY STOPPED IMPACTING TIME AND COST.
FISCAL RISK	THE PROJECT COULD OVERSPEND AND RUN OUT OF AVAILABLE FINANCE.	THE PROJECT MIGHT HAVE TO BORROW ADDITIONAL FUNDS AT A HIGHER INTEREST RATE IMPACTING COSTS.
SOCIETAL RISK	SOCIETY COULD BE IMPACTED THROUGH A PROJECT ERROR SUCH AS THE LEAKING OF TOXIC GASES.	HERE THE PROJECT WOULD MOST LIKELY BE BOTH CLOSED AND SUED PUTTING TIME, COST AND THE QUALITY OF PROJECT DELIVERABLES AT RISK.
ENVIRONMENTAL RISK	A PROJECT COULD IMPACT THE BIO-DIVERSITY OF AN AREA.	IF THIS WERE DONE THOUGHTFULLY IT COULD ADD BENEFIT TO THE PROJECT. THE REVERSE IS OF COURSE WOULD ALSO BE TRUE.
CONTRACTUAL RISK	THE PROJECT MIGHT DELIVER LATE DUE TO POOR CONTRACT MANAGEMENT.	PENALTY PAYMENTS COULD RESULT.
STAKEHOLDER RISK	POOR COMMUNICATION WITH STAKEHOLDERS COULD LEAD TO A LACK OF DEFINITION IN THE PROJECT DELIVERABLES.	THE PROJECT MAY NOT DELIVER THE BENEFITS REQUIRED, OR MAY NEED TO REWORK THE DELIVERABLES IMPACTING TIME, COST AND POTENTIALLY QUALITY OF DELIVERABLES.
NATURAL RISK	THE PROJECT COULD BE IMPACTED BY FLOODS, SEVERE STORMS OR OTHER UNUSUAL NATURAL EVENTS.	THE IMPACT OF THESE ON THE PROJECT COULD BE DELAY AND ADDITIONAL PROJECT SPEND.
PROJECT RISK	POOR PROJECT PROCEDURES, OR LACK OF EXPERIENCE OF KEY PROJECT PERSONNEL, COULD LEAD TO POOR MONITORING AND CONTROL OF THE PROJECT.	POOR MONITORING AND CONTROL WILL IMPACT THE SMOOTH RUNNING OF THE PROJECT RESULTING IN SUB-OPTIMAL BUSINESS CASE DELIVERY.

(Continued)

Table 9.1 (Continued)

POSSIBLE RISK SOURCE	*POSSIBLE ISSUE*	*POSSIBLE PROJECT IMPACT*
STRATEGIC RISK	A NEW STRATEGIC REVIEW WITHIN THE ORGANISATION COULD MEAN THAT THE PROJECT IS NOW IN DANGER OF ADVERSELY IMPACTING THE ORGANISATION'S STRATEGIC DIRECTION.	IF THE PROJECT IS NO LONGER STRATEGICALLY ALIGNED, IT COULD COST MONEY WITHOUT DELIVERING BENEFITS.
REPUTATIONAL RISK	AS A RESULT OF A PROJECT DELIVERING A POOR (OR GOOD) PRODUCT AND THAT PRODUCT BEING SOLD, THE ORGANISATION'S REPUTATIONAL STANDING COULD BE IMPACTED.	IF THE PRODUCT IS POOR, THEN THE REPUTATION OF THE ORGANISATION WILL SUFFER MEANING THAT SALES REVENUE IS LIKELY TO FALL. THE REVERSE COULD OF COURSE BE TRUE IF THE PRODUCT WERE GOOD.
PRODUCT RISK	THE PROJECT OUTPUT MAY NOT MEET REQUIREMENTS OR IT MAY BE FAULTY.	THIS COULD ADVERSLY EFFECT THE RETURN ON THE PROJECT, POTENTIALLY PUTTING THE ORGANISATION'S FUTURE AT RISK.
GOVERNANCE RISK	THE PROJECT MANAGEMENT PROCESSES IN PLACE, PERHAPS AT AN ORGANISATIONAL RATHER THAN PROJECT LEVEL, ARE INADEQUATE.	THE PROJECT WILL NOT BE DELIVERED AS EFFICIENTLY NOR AS EFFECTIVELY AS IT MIGHT HAVE BEEN.

9.2.1 *How Can Risk Be managed?*

Risk has been defined above as the probability multiplied by the impact of an event happening. Risk Management can be defined as 'a process that allows individual risk events and overall risk to be understood and proactively managed, optimising success by minimising threats and maximising opportunities.'[2]

Some important points to note from this definition are that risks should not be looked at in isolation. A risk could be identified that a particular component in use on the project might fail during the next week. This would attract a measure of probability and impact and might appear as a key risk for the project. If, however, it were analysed as a risk that the component might fail on either Monday, Tuesday, Wednesday etc. all the way to Sunday, then the risk likelihood would presumably be reduced by a factor of seven. The overall risk to the project however, would remain unchanged and the seven 'individual risk events' should be treated as one 'overall risk' event recognising the commonality within the source of the risk and hence in the required mitigating action.

Proactive management is key since the closer the project gets to a risk occurring, there is generally less that can be done to affect the outcome and in addition the cost of taking any action will normally increase. Whether or not proactive action can be taken is dependent on the correct identification of risks – if a risk is not identified it can obviously not be proactively managed.

Given the definition of risk as probability times impact, the use of the word 'impact' does not dictate whether the impact of the risk is either a negative or positive influence on the project outcomes. The potential negatives impacts are often referred to as 'threats' and the possible positive impacts as 'opportunities'.

Risk management within a project is a continual process. Risks can evolve at any point within the project lifecycle. Outside of the project world, the need to manage risks is widespread and a number of different processes have been defined. One model based upon that in HM Treasury's Orange Book[5] is outlined below in Figure 9.2.

The model has four stages: identify the risks, assess the risk, address the risk and review the impact of the risk treatment. The continuous nature of risk management is clearly shown by the circular nature of the diagram.

The APM take a very similar approach to risk management. Their Risk Management Model[6] is similar to that of the PMI (Project Management Institute) and it closely reflects the Treasury model at Figure 9.2.

The main difference of course between the Treasury and the two Project Management bodies is that the Treasury is a permanent organisation, while the work delivered under the auspices of both the APM and PMI is transitory in nature. This means that in order to provide a risk management approach, the APM and PMI both need to establish the rules for managing risk for each and every new project. A fifth step is therefore required. The fifth step is the initiation, or setting up, of the risk management process as shown at the beginning of Figure 9.3 below.

The presentation in Figure 9.3 appears less circular than Figure 9.2 at first glance, but on closer inspection, it shows the ability to move backwards and forwards between the phases as necessary, and also implies that the process is a through life process showing a link from the end of the 'Review Risks' phase back to the 'Identify phase'

Each phase has a different focus.

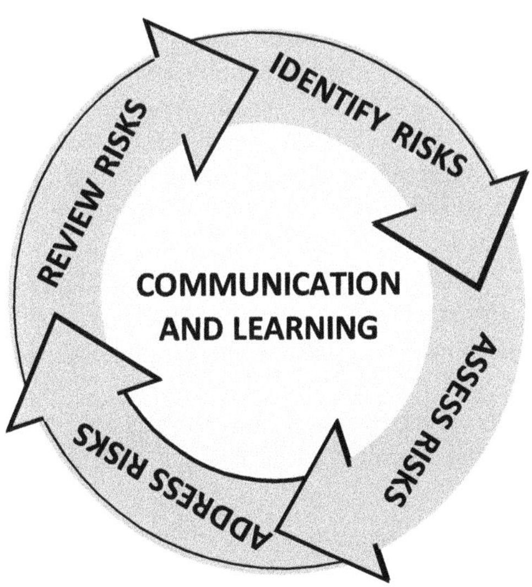

Figure 9.2 – a Risk Management model based upon the HM Treasury model.

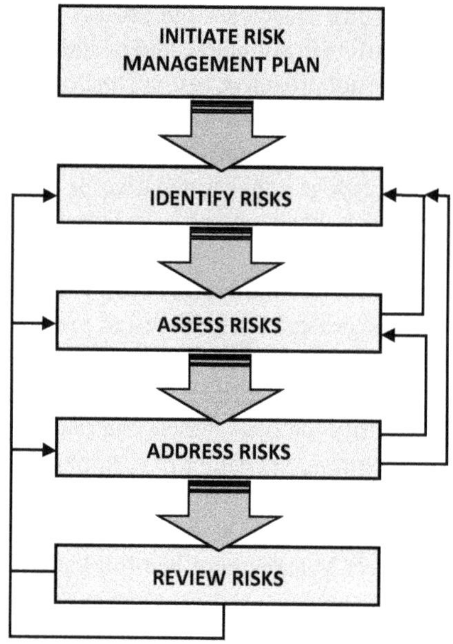

Figure 9.3 – a project Risk Management model based upon the APM and PMI models.

9.2.1.1 *Risk Management – The Initiate Phase*

The initiate phase sets out the way in which the project team will manage the risks. For example, how will the project run the remaining four phases of the process? What is the project's attitude to risk, is it risk tolerant or risk adverse, are there areas of risk which the project considers more harmful than others? The initiate phase is also required to identify the key roles within the risk management process. If a potential risk has an impact of £1,000 then who can sign-off the mitigation for this? Is this different if the potential impact is £10,000? £100,000? Presumably in the case of the larger sums, the approval will need to come from a higher authority, for example, to be escalated to the project manager or project sponsor rather than the workstream leader. Any costs relating to risk mitigation actions will need to be assessed and authorised in a similar way. A risk mitigation might cost £25,000. Is it worth spending this amount if the impact on the project, should the risk occur would be £250,000? What if the probability of the risk occurring was 1%? What if the probability of the risk occurring was 75%? Possibly the two answers would be different, but the approach to this question should be set out and agreed in the Risk Management Plan established as part of the initiate phase.

The Risk Management Plan will need to agree a raft of different things, not just the way in which risks are assessed, measured and managed within the project. It will also need to consider which tools, techniques, methods, processes and criteria are best suited to managing the project's risk.

All of the above need to be agreed by the project team and evidenced in the Risk Management Plan, documented as part of the overall Project Management Plan.

9.2.1.2 *Risk Management – The Identify Phase*

As considered in Section 9.2, risk can come from many different sources both inside and external to a project. Identifying all the major risks, let alone the lesser ones, can therefore be difficult.

Regular team meetings are a vital part of any project and risks will be identified in these, but to leave it to chance as part of these meetings would be brave to say the least. A more focused approach to risk identification is required. A risk investigation meeting is often held at the beginning of a project where various project scenarios are discussed. Those in attendance should include as many of the team and stakeholders as feasible. It may be worthwhile holding more than one meeting in order to broaden the base in terms of risk identification. A common way of running these meetings is to begin with a discussion of the project outline and deliverables before moving onto a brainstorming phase where nothing raised is discouraged or ignored. This enables the project manager to explore the widest possible set of views on the project and to also identify particular contributors who have an area of expertise relevant to the project. These may be people who have worked on similar projects in the past or who have in-depth knowledge or skills relevant to the project as a whole, or to significant parts of it. These individuals can be interviewed in greater depth by the project team in order to gain maximum benefit from their previous experience of comparable projects.

In order to help inform these meetings, a review of the projects record office can provide valuable input. Records from previous projects can potentially detail where similar projects have experienced risks that could be common to the current project. Post Project Review reports are another good source of risk related information as are checklists based on previous projects delivered. It is also worth discussing, at these meetings, any assumptions which the project may have made and ensuring that they are valid.

9.2.1.3 Risk Management - The Assess Phase

Once the project risks have been identified, the importance or potential impact on the project of each risk should be assessed. One of the most straight forward ways to do this, is to refer back to the definition of risk given at the beginning of Section 9.2. If risk = probability x impact, then it should be possible to calculate, or estimate, the value of both the probability and the impact attributable to each risk.

Once these figures are calculated, they can be mapped onto a probability and impact grid such as the one below at Figure 9.4.

The positioning of the risk when plotted on the grid represents a higher level of risk as the plot moves from bottom left to top right, for instance risk A is clearly lower than risk B or C. Whether C is a larger risk than B however will depend upon the values attributed to a low, medium or high probability compared to a low, medium or high impact. Each project can assign whatever values seem reasonable to these variables, and these values will be recorded in the Risk Management Plan, as described in Section 9.2.1.1. If equal numerical values are assigned to both the low, medium and high ratings of probability and impact, then the project would demonstrate that it placed equal importance on both factors. If however, numerical values were assigned as per Figure 9.5, then the project could express a view as to what was of the greater concern – the probability of a risk occurring or its impact should it occur.

In this instance, Risk A would have a value of 2 x 1 = 2, Risk B of 5 x 3 = 15 and Risk C of 10 x 2 = 20. This would suggest that risk C should be the first that should be treated as having been assessed, it has the highest risk value.

Section 9.2.1 raises the interesting case of commonality between sources of risk and states that individual risk events which are directly related to each other should be treated as one overall risk when effective risk mitigation can be achieved through one and the same action.

Risks can also be related in other ways which should be considered. Secondary risk is such a case. If one project risk happens, it can be that the probability of a second risk occurring

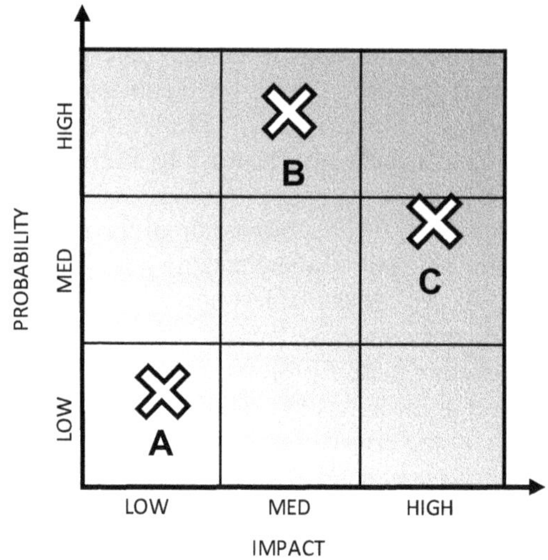

Figure 9.4 – a probability impact grid.

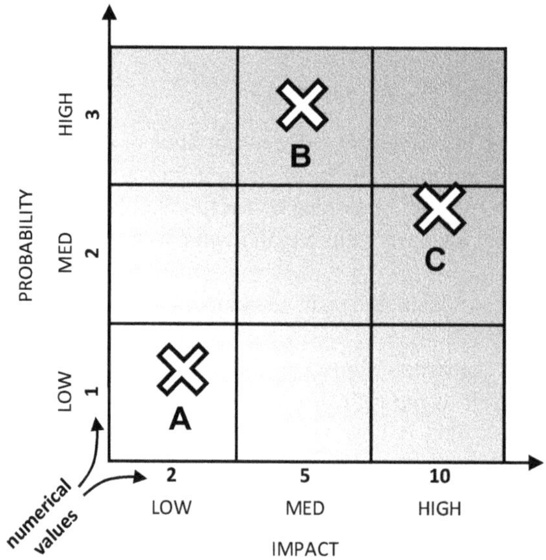

Figure 9.5 – a probability impact grid – with numerical values assigned.

changes and becomes more likely to occur, this is known as a secondary risk. Where secondary risks occur, they should not be treated as individual risk events but more as 'compound risk' events. A very simple example might be a construction project where four walls have been built but the mortar has not properly cured. One wall might, perhaps due to having more openings, be more likely to collapse in a storm than the other three. However, if it does, then the possibility, or risk, of one of the remaining walls falling would increase due to a loss of structural integrity. This is a 'secondary risk' and often in projects the project manager needs to be aware of which risks could materialise given the occurrence of an initial risk. These secondary risks can be treated in the same way as those outlined above and their probability mapped onto a probability impact grid.

9.2.1.4 *Risk Management - The Address Phase*

Once the risks have all been ranked as described in Section 9.2.1.3, mitigation actions need to be planned and implemented. When planning mitigation action, the cost of mitigation has to be considered. In Figure 9.5 it appears as though risk C must be mitigated, but if as would be normal, the project has a limited risk mitigation budget, risk A should perhaps remain unmitigated. The one factor that has not been considered so far though is the cost of mitigation. If the impact of C on the project was £2m but the only way of mitigating it involved constructing an exact replica of the component at a cost of £2.5m, then, from a financial point of view, it would be best to not mitigate the risk but instead to watch it carefully and to determine if anything changes the impact or mitigation cost figures. Risk A on the other hand, with an assumed impact of £50,000 should it occur, might be very cheap to mitigate, potentially it might be fully mitigated for £100. This would result in mitigation action being taken for risk A but not risk C, which based solely on Figure 9.5 and ignoring the cost of mitigation, would be counter intuitive. This means that before any risk mitigation can be put into place, the cost implications need to be evaluated.

Risks can be mitigated in a number of different ways and by a variety of different types of action. These options lie on a spectrum between doing nothing, other than watching the risk profile, to fully alleviating the risk and will obviously come with associated project costs.

The range of possible risk treatments are described below.

The first approach a project manager can take is to redesign the way in which a project is delivered to ensure that the risk can no longer happen. This could be achieved in a number of ways depending on the type of risk being faced. If there was a risk that an important component coming from abroad by ship might be delayed due to industrial action at one of the ports it has to pass through, then the risk of delay could be fully mitigated by sending it by air. If the risk was that a particular contractor, although cheap, might not be able to deliver satisfactorily, then another firm with a better reputation could be employed at extra cost meaning that the risk of non-performance from the original contractor was removed. The taking of an alternative approach is a way of avoiding the risk – the first of four risk mitigation strategies.

The second approach to mitigating risk would be to reduce the degree of risk so that it was at an acceptable level for the project. This would mean that either the probability of the risk occurring or the impact of it would need to be reduced. In the case of providing footings for a building constructed in a new environment where the company constructing it had little experience of local ground conditions, it might be unclear, within the geographic conditions, whether or not the footings would be strong enough to support the planned construction. To reduce the risk to an acceptable level, the footings could be built with additional reinforcement which would mean that while the footings could still fail, the chances of this happening had been reduced to an acceptable level.

A third approach that could be taken in the example above, would be to sub-contract the footings to a specialist company. The contract written with them for this work would then reflect the share of risk between the two parties. The contract could be written in such a way that if the footings did fail, then financial recompense would be paid to the company contracting with the specialist such that it balanced the impact of the risk. In this case the impact of the risk occur becomes neutral, it is fully treated and sits entirely with the third party specialist company. In terms of risk mitigation actions, this is known as 'risk transfer'.

STRATEGY:	ACTION:	RESULT:
AVOID	select an alternative approach	risk is no longer present
REDUCE	redesign to lessen the risk	risk is at an acceptable level
TRANSFER	pass management of risk to a 3rd party	risk is with someone else
ACCEPT	do nothing, but review periodically	proactive management required

Figure 9.6 – summary of risk mitigation actions for negative risks (threats).

STRATEGY:	ACTION:	RESULT:
EXPLOIT	select an alternative approach	risk IS present
ENHANCE	redesign to increase the risk	risk is MORE likely to occur
SHARE	pass management of risk to a 3rd party	opportunity is with someone else
REJECT	do nothing, but review periodically	proactive management required

Figure 9.7 – summary of risk mitigation actions for positive risks (opportunities).

Some risks may be too expensive, or just too small in terms of impact, or to unlikely to happen that the project manager may choose simply to accept the risk and to regularly review it to ensure that the risk profile doesn't significantly change. Accepting a risk is the fourth possible approach to mitigating it.

Reviewing Figure 9.6 reveals that only negative risks or threats have been considered. Risks can, as already discussed be positive. Figure 9.7 shows what are in effect mirror images of the results of strategies shown in Figure 9.6.

9.2.1.5 Risk Management – The Review Phase

The review phase of the risk management phase needs to collect the identified risks into one place and then record the actions to be taken, or taken, in order to mitigate the risks. This is often done through use of a document known as a Risk Log or Risk Register. Two things are important in this register – a comprehensive list of risks being managed and the decisions that have been made in order to mitigate these risks. The headings for the Risk Register can appear as shown below.

The columns in Figure 9.8 are each numbered.

Column 1 is merely a project risk reference number.

Column 2 relates to the identify phase of the risk management process. Each risk should be described as fully as possible. What on the surface appears a simple risk such as loss of power to a site may need to be addressed by one of two distinctly different mitigation actions that will only become clear if the risk is fully described. If the risk of 'loss of power to the project site' was the risk entered into Column 2, the mitigating action entered in Column 7 could, for example, be 'stockpile increased diesel supplies for the generators' or it could be 'protect buried power cables from site machinery'. Knowing which one is the correct mitigation would depend

PROJECT NAME:							PROJECT MANAGER:					
VERSION:							DATE:					
1	2	3	4	5	6	7	8	9	10	11	12	13
Ref No.	Risk Description:	Probability H, M, L	Impact H, M, L	Overall risk rating (cols. 3 x 4)	Date Raised	Review Date	Planned Action to Mitigate Risk:	Date Mitigation Completed	Risk Owner	Steering Committee Owner	Mitigation action effective Y/N	Date Closed

Figure 9.8 – the headings within a Project Risk Register.

on the nature of the risk being fully specified. If for instance, the risk description said 'loss of power to the project site due to buried power cables being cut during ground works' a far more suitable and targeted mitigation could be considered than if the risk description was simply 'loss of power to the project site'.

Columns 3,4 and 5 take their information straight from the assess phase of the risk management and will result in the figure in the Column 5 (calculated by multiplying Column 3 by Column 4,) that represents the overall size of risk presented to the project.

Columns 6 and 7 are important as they will highlight when each risk was initially recognised and how it is being managed and reviewed, providing evidence of ongoing risk management.

The second half of the table, Columns 8–13, focuses on the mitigation action itself. Columns 8 and 9 record the mitigation action agreed upon by the project team, or how the risk has been addressed. This takes the form of one of the options shown in Figures 9.6 and 9.7 and includes a detailed description of the mitigation action specific to the risk being addressed. Column 9 keeps track of whether the approved mitigation action has actually been completed.

What is equally important, if not more so, than identifying a risk mitigation action is to identify the person who will be responsible for ensuring the mitigation action is completed in a timely fashion. Column 10 should identify this person. Column 11 indicates which of the projects senior owners should oversea that the mitigation is successfully delivered. Columns 12 and 13 represent the review phase of Figure 9.3, overseeing the effectiveness of the mitigation once the action has been completed, and allows the risk to be closed. If the time period in which the risk can occur has passed, then the risk can also be recorded as closed.

New project risks can appear at any time. The risk log and associated risk management process are therefore 'living' activities which must be enacted throughout the project lifecycle, hence the circular nature of risk management discussed in Section 9.2.1. The risk management process as a whole is focused on the protection of the project benefits identified in the business case (see Chapter 2) and to the maximisation of the gap between the project's costs and its benefits. Without this management of risk, projects can fail drastically as illustrated in the following case study.

9.3 Project Review: The impact of Risk Management on the Project: the Space Shuttle

9.3.1 *The Project Background*

The Space Shuttle Challenger 51-L was part of NASA's STS (Space Transportation System) programme.[7] The flight itself was the 25th mission within the programme which aimed to enable travel into space using a system of reusable spacecraft. The Space Shuttle Programme itself was approved in 1972 by the then President of the United States,[8] Richard Nixon. The primary aim of the programme was to enable routine and economical flights into space, and it had its roots in 1967, when the President's Science Advisory Committee supported the idea of a reusable spacecraft by recommending that studies be made "of more economical ferrying systems, presumably involving partial or total recovery and re-use."[9]

Fourteen years later, the first launch of the STS Programme saw Columbia take off on April 12, 1981, and the programme continued until the successful landing of the final flight on July 21, 2011. In all NASA's Space Shuttle fleet flew 135 missions and helped to construct the International Space Station.[10] One flight, Challenger 51-L's, however suffered a terrible disaster.

The purpose of Challenger 51-L's flight[11] was to launch both a tracking and data relay satellite and the Spartan Halley's Comet observer into space. Notably one of the crew was on-board in order to provide live educational broadcasts with the intention of beaming them to classrooms around the world. Other aims of the flight included the launch of the second Tracking and Data Relay System (TDRS) satellite which once deployed would enable near-continuous communications during future shuttle missions. The deployment of the Spartan Halley astronomy satellite focused, in partnership with several international spacecraft, on providing observations of Halley's Comet, which was due to pass close by during its 76-year orbit around the sun. The satellite would be launched on the third mission day and retrieved two days later after its the observations had been completed.

It was the two live educational broadcasts planned for the mission's sixth day which were causing the most interest. The focus of the first broadcast called "The Ultimate Field Trip," was to compare life on the Space Shuttle with that on Earth. The second called "Where We're Going, Where We've Been, Why?" was to concentrate on exploring the importance of conducting research in space.

9.3.2 *The Project Issue*

The problem was catastrophic. On the 28th of January 1986, NASA launched the Space Shuttle Challenger. At 4.30 pm Challenger lifted off with a crew of seven on-board. Just seventy-three seconds later, the Space Shuttle's external fuel tank collapsed. The fuel it was carrying ignited causing a fireball resulting in Challenger exploding causing the death of all seven on-board.[12]

The launch window was determined principally by the need to provide the best possible conditions to make the observations of Halley's Comet. The Level 1 Flight Readiness Review was designed to ensure that all aspects of flight preparation giving rise to any question were fully addressed. The Level I Flight Readiness Review for the mission took place on January 15, 1986, and no outstanding concerns were registered.[9]

The launch was designed to send two Solid Rocket Boosters plus the orbiter, Challenger, into spaced. The purpose of the Rocket Boosters was to enable the craft to break free of the Earth's atmosphere before establishing its orbit. Shortly after launch an explosive burn of the hydrogen and oxygen propellants in the booters destroyed the External Tank and exposed the orbiter to severe aerodynamic loads that caused its complete structural breakup. All seven crew members perished. The two Solid Rocket Boosters flew out of the fireball and were destroyed by the Air Force range safety officer 110 seconds after launch.[8]

The Rogers Commission Report[9] went on to say that repeated puffs of black smoke were visible from one of the boosters and that this indicated that the seals within a joint were faulty. The black colour and dense composition of the smoke puffs suggest that the grease, joint insulation and rubber O-rings in the joint seal were being burned and eroded by the hot propellant gases and that this contributed to the accident. The Rogers Commission Report[9] stated that 'at the cold launch temperature experienced, the O-ring would be very slow in returning to its normal rounded shape....Thus, it is probable the O-ring would not be pressure actuated to seal the gap in time to preclude joint failure due to blow-by and erosion from hot combustion gases.'

This led to the destruction of the seals (linked to the O-rings) that were intended to prevent hot gases from leaking through the joint during the propellant burn of the rocket motor.[9]

9.3.3 How Could a Developed Risk Management Process Have Helped?

While a disaster like this is never due to one factor alone and it would be wrong to suggest that the management of risk was solely responsible for this tragedy, it is however, of interest to review what official reports and academic papers have had to say.

The disaster sparked a number of official papers, notable amongst them were the report prepared in the US by the Committee on Shuttle Criticality Review and Hazard Analysis Audit of the Aeronautics and Sace Engineering Board,[8] (CSCR), and the Rogers Commission Report commissioned by President Regan.[9] Completed in June of 1986, one of the recommendations of the Rogers Commission, noted by the CSCR, was to 'identify those items that must be improved prior to flight to ensure mission success and flight safety.'[9] Indeed, it was the report of the Rogers Commission which recommended and led to the establishment of the CSCR in order to 'verify the adequacy of the effort' employed to identify these items.[8] The CSCR recognised that projects such as the Challenger 51-L could only be attempted in the face of risk, as project managers would agree, noting that all projects have, by definition, an element of risk within them. The CSCR went on to say however, that in order help ensure the best outcome, any such project should include what today we would recognise as some of the basics of a risk management process consisting of:

- A comprehensive method for identifying project risk
- A specific, quantitative method for assessing or estimating the safety risks of the system
- A system of risk management to ensure safety risks are at a level acceptable to the approval authority. This would include:
    - Establishing acceptable risk levels
    - Implementation of actions to achieve acceptable risk levels
    - Approval and certification of the system
    - Continued quality assurance of the system.

The steps included here, recommended by the CSCR are very similar to those discussed in Figure 9.3.

Rickard[12] and Allen et al.,[13] make some interesting observations in relation to the process outlined above and the accident. One of the key factors identified was that, as the Rogers Commission determined, three exceptionally cold nights prior to the shuttle launch had caused the O-rings to become brittle. The approach taken was that where 'unanticipated problems continue to occur without significant issues' the notion of acceptable deviance was established. This approach has been summed up[14] by saying that 'when playing Russian roulette, the fact that the first shot got off safely is little comfort for the next.'

In assessing the weakness of the risk management process, it should be noted that the danger of O-ring failure was known prior to the launch. Rickard[12] reports that the engineers from the manufacturers of the O-rings, the Thiokol Chemical Corporation, 'voiced their concerns to NASA regarding the potential non-performance of the O-ring due to abnormally low temperatures.' Both NASA and Thiokol knew that there was room for improvement within the seals, how much room was less clear as the Thiokol engineers expressed the view that their own testing might be flawed due to software issues. A definitive recommendation with regard to the launch, was required and during a conference call, the engineers recommended postponing launch for a few days until temperatures rose. This would have had the potential to put the rendezvous with Halley's Comet at risk. The conference call ended. Thiokol management spoke to the chief engineer and asked him to 'take of his engineer cap and put on his management cap' when making his recommendation. The suggestion was that he if he didn't he might be viewed as not giving sufficient prioritisation to the overall organisational goals.

9.3.4 Conclusions on the Space Shuttle

In terms of the risk management process, had it been applied correctly?

The risk of the O-ring failure had been correctly identified. Step 1.

The assessment of the risk, Step 2, however had been compromised. Undue influence had been exerted and the Thiokol engineers had been 'encouraged' to change their minds without the necessary due diligence, and the level of risk had been incorrectly assessed. As the CSCR recommendation made clear, an improved risk management process would have helped prevent this. The CSCR report also highlighted the need for a formalised risk review process, which perhaps hints at the existence of risks that weren't recognised at Step 1, for example the risk of political pressure and the temporal risk due to the close proximity of Halley's Comet. A full Risk Management Plan detailing points such as acceptable risk levels, necessary actions to achieve acceptable risk levels and the approval, certification and quality assurance of the risk management system should have been established at the outset of the project so that in the heat of the moment any deviance from this would have needed to be fully explained and approved rather than allowed to happen without due diligence.

9.4 Risk Management Before the Birth of Modern Project Management*

(* In this section, the research relating to the construction of HMS Barham 1912–1915 mentioned in the introduction is drawn upon to help illustrate how the subject of this chapter has developed.)

In this, the review of Project Management practices before the establishment of what is termed 'Modern Project Management', attention once more is focused on the construction of HMS Barham (1912–1915) and the review of the project records that have survived. It has been noted before that in searching for historical records the researcher is at the mercy of what is considered important and hence what should be kept by the archivist.

The records of risk management have unfortunately not fared well in respect of HMS Barham. What cannot be seen directly, is evidence of the four steps of identify – assess – address – review, that Figure 9.3 would lead us to expect. Instead, what can be observed is conformation of the impact of these steps. One example of this, relating more to the 'address' step, is shown in Photo 9.1 below.

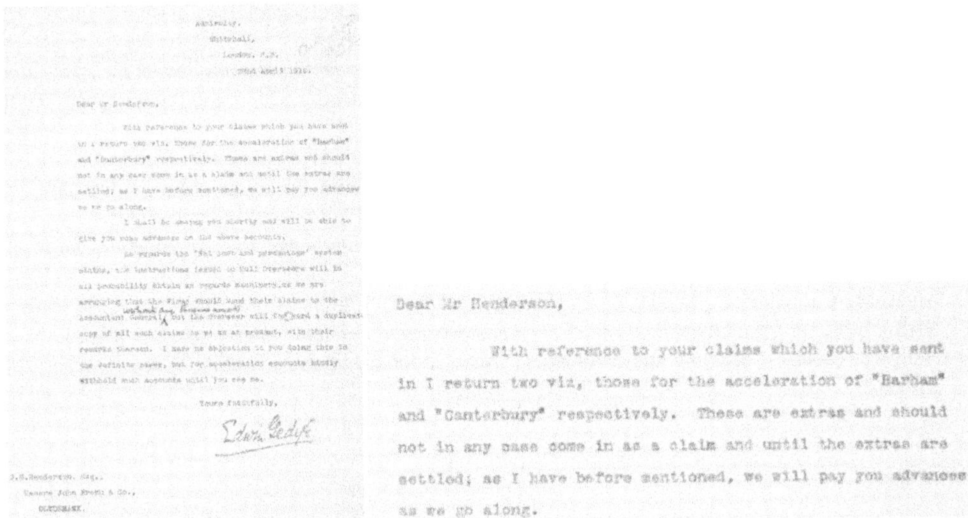

Dear Mr Henderson,

With reference to your claims which you have sent in I return two viz, those for the acceleration of "Barham" and "Canterbury" respectively. These are extras and should not in any case come in as a claim and until the extras are settled; as I have before mentioned, we will pay you advances as we go along.

Photo 9.1 – A letter dated 22/4/1915 to John Brown & Co Ltd relating to 'acceleration' payments. (With enlarged extract on the right-hand side.) By permission of the National Records of Scotland and University of Glasgow Archives & Special Collections, Records of Upper Clyde Shipbuilders Ltd, Clydebank Division collection, GB248 UCS 1/21/9.

The letter above discusses Admiralty payments for 'the acceleration' of work on HMS Barham and could claim to be evidence of risk management since although it shows that the need for 'acceleration' was considered to be 'extras', the need for 'acceleration' was both identified and addressed. The letter addresses the risk of late delivery and proposes a costed solution and payment terms. In order to have reached the conclusions in the letter, the risk of late completion due to a lack of labour must have been identified and assessed by the John Brown Shipyard before being addressed. Photo 9.1 therefore evidences the first three stages of the 'identify – assess – address – review' Risk Management model established in Figure 9.3.

References

1 Project Management Institute. (2021). *The standard for project management and a guide to the project management body of knowledge* (7th ed.). Project Management Institute, Inc.

2 APM. (2019). *APM body of knowledge* (7th ed.). Association for Project Management.

3 Ohara, S. (2017) *P2M – A guidebook of project & program management for enterprise innovation* (3rd ed.). Project Management Association of Japan (PMAJ).

4 Wilkinson, I. (2023, February 7). Trains in Spain stuck mainly on the plain after tunnels blunder. The Times. https://www.thetimes.co.uk/article/new-trains-don-t-fit-through-tunnels-spanish-ministry-of-transport-admits-mgvqzc6b9 [Accessed 18 March 2024].

5 UK Government. (2023). *The orange book management of risk – Principles and concepts*. https://assets.publishing.service.gov.uk/media/6453acadc33b460012f5e6b8/HMT_Orange_Book_May_2023.pdf [Accessed 22 March 2024].

6 APM. (2012). *APM body of knowledge* (6th ed.). Association for Project Management.

7 Forrest, J. (1986). *The space shuttle challenger disaster*. Metropolitan State College.

8 Slay, A., & National Research Council. (1988). *Post-challenger evaluation of space shuttle risk assessment and management*. National Academies.
9 NASA. (1986, January 31). *Report to the president by the presidential commission on the space shuttle challenger accident.* https://sma.nasa.gov/SignificantIncidents/assets/rogers_commission_report.pdf [Accessed 20 March 2024].
10 NASA. (2024). *Space shuttle*. https://www.nasa.gov/space-shuttle/ [Accessed 20 March 2024].
11 NASA. (2024). *35 years ago: Remembering challenger and her crew*. https://www.nasa.gov/history/35-years-ago-remembering-challenger-and-her-crew/ [Accessed 20 March 2024].
12 Rickard, K. (2015). *Risk management and the space shuttle challenger disaster*. Institute of Public Administration.
13 Allen, M., Carpenter, C., Dydak, T., & Harkins, K. (2016). Causes of project failure: Case study of NASA space shuttle challenger. *Journal of Engineering and Economic Development*, *3*(2), 23.
14 Hall, J.L. (2003). Columbia and challenger: Organizational failure at NASA. *Space Policy*, *19*(4), 239–247.

Chapter 10

The Project Budget and Cashflow

10.1 The Purpose of the Project Budget

The APM[1] define cost planning and control as 'the estimation of costs, the setting of an agreed budget and the management of actual and forecast costs against a budget'. The Project Budget is therefore a document which once established and agreed amongst the project team and project sponsor must be constantly reviewed to ensure that the project remains on track financially and that any changes in scope are accurately reflected. The PMI[2] link the budget back to the estimating processes, discussed in Chapter 7, and see the budget as evolving from these estimation procedures. The key aspect of the PMI definition of a budget is that it is an approved estimate for a project or WBS components or activities on the project schedule. If the budget is 'approved' then this indicates a degree of consideration and consultation linked with the budget before it is signed off as agreed which helps to validate and test the budget, both for accuracy and completeness. Having defined the project budget, the PMI offer a definition of the Cost Baseline. This they define as 'the approved version of the time-phased project budget, excluding any management reserves, which can be changed only through formal change control procedures and is used as a basis for comparison to actual results'.[2] Thus, two additional aspects are introduced to the definition of the budget - the time-phasing of the project costs and the need to use it as a project baseline which can only be adjusted through a formal process.

In the same way that the PMI makes the connection between the scope (Chapter 8) and the budget, P2M[3] reflects on the importance of scope development as a precursor to developing the budget stating that a 'poorly defined scope may cause confusion in the project and result in changes or reworks, thereby exceeding the budget'. Scope definition is an obvious and necessary precursor to the development pf a project budget, beyond this though P2M also discusses the 'three operation process of cost management'. This ties together the three processes of cost estimation, budget setting and monitoring and control of the budget, showing the connections and interrelation between the tools and techniques discussed in some of the other chapters of this book (Figure 10.1).

Throughout this process, budget management is needed in order to cope with and control variables such as changes in economic conditions, agreed changes to project scope, costs resulting from changes in product specifications and quantities as well as managing payments to suppliers.[3]

10.2 How Is the Project Budget Delivered, Managed and Used?

There are a variety of terms associated with budgeting and cashflow management which are used to describe the numerous types of payment which are made through the life of the project.

DOI: 10.1201/9781003470748-11

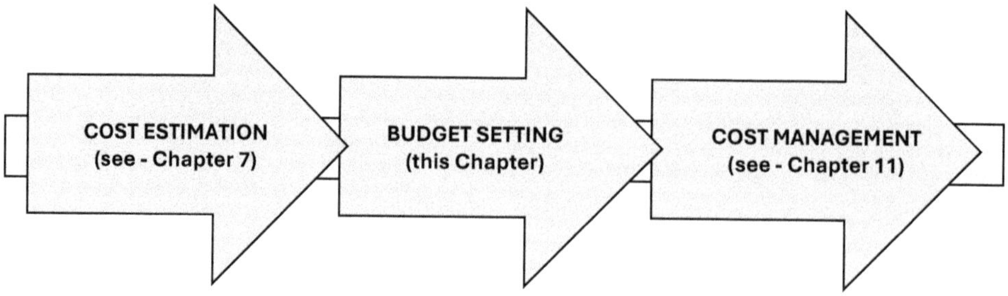

Figure 10.1 The 'Three Operation Processes of Cost Management' (P2M[3]).

Terms such as budget cost and budget commitment are expressions more often used in the planning of a project's expenditure whilst phrases such as accrual, cash flow and forecast are, arguably, more associated with the monitoring of the budget. Terms such as out-turn costs, actual costs and sunk costs are perhaps used more within the reporting of the project spend. The variety of terms does however serve to illustrate the key purposes and aim of the project budget – to help in the planning, and in the monitoring of the project as it delivers, and to give a measure of project success from a financial reporting standpoint.

The importance of controlling the project budget can be seen diagrammatically in Figure 10.2.

One of the interesting things to note in relation to the curves is their shape. They will almost invariably be what is called an 'S' curve. All the curves in Figure 10.2 are 'S' curves albeit shallow 'S's. The reason that project cost curves will be this shape is that at the start of the project planning is paramount and resource consumption is relatively low – nothing is being physically constructed, hence the gradient of the curve, or increase in total cost over each time period is comparatively low. As the project reaches the manufacture stage resource consumption increases reflected in the curve which has a steeper gradient in its middle section. Once the project nears completion the rate of resource consumption flattens out once more giving the

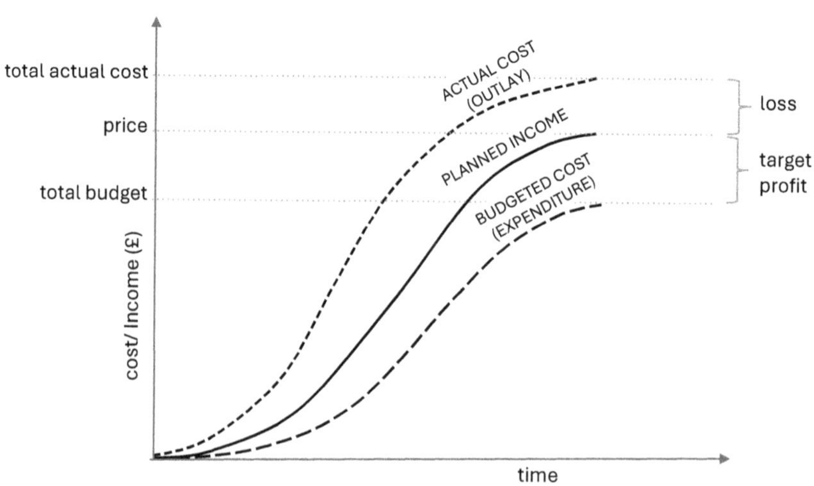

Figure 10.2 The project budget income and cost curves.

curves their distinctive 'S' shape as various resources are stood down an resource consumption generally falls.

The relative positions of the project budget income and cost curves will of course vary depending on the financial outcome of the project. As drawn in Figure 10.2, the budgeted cost is less at the end of the project than the planned income, so a project profit is expected, shown as the difference between the two. However, the Actual Cost or project outlay is shown as being substantially larger than the income and so due to a lack of control over the cashflow the project is showing a loss at the end of the project. This shows the importance of both accurate budgeting and budgetary control through the project lifecycle. Once agreed the budget can help to establish the project profit margin, help determine how much finance needs to be raised and to also act as a baseline providing a basis for financial control of the project.

10.2.1 Creating the Project Budget

Having agreed the project scope, the next items to consider are the resources that the project will need to procure and which will directly contribute to the project costs. These costs can of course be many and varied,but typically would include costs relating to manpower, materials, equipment, overheads, sub-contractor costs and even sums set aside for any potential contingency. It may not be the project managers responsibility to put the budget together, but it is certainly his job to understand it, to monitor and to deliver it.

The project budget requires the project to have been scoped so having studied this and defined the necessary resources, the construction of a time-phased budget can begin. Earlier, in Chapter 8, a project Gantt Chart was derived showing the tasks involved in the restoration of a car, it is shown again at Figure 10.3.

This Gantt Chart can be used to base a phased budget on.

Suppose the work involved consisted simply of labour, materials and a project management overhead. The project management overhead could consist of a number of sub-charges and

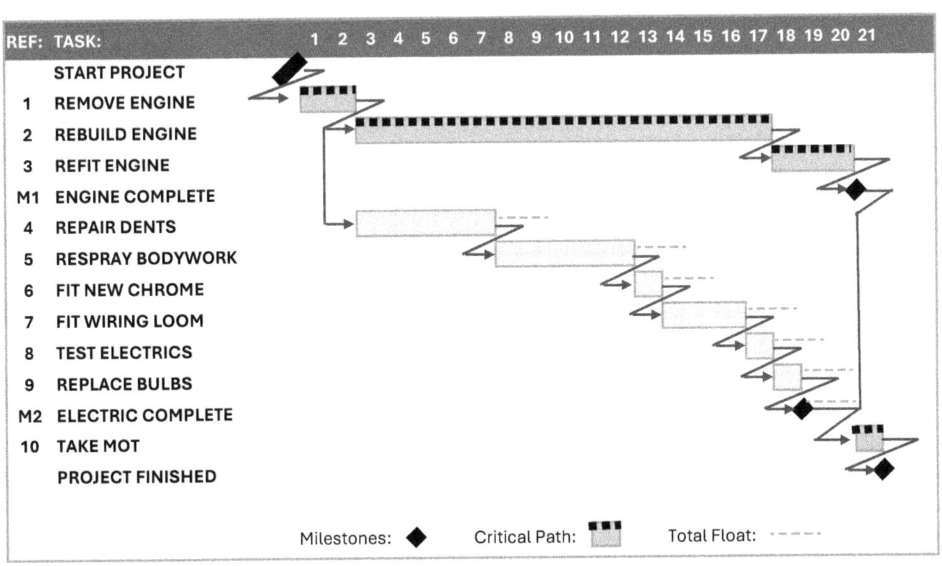

Figure 10.3 Project chart as originally shown in Figure 8.15.

Table 10.1 Estimated Project Costs

Task No.	*Task Description*	*Task Duration (Days)*	*Labour Units per Day**	*Materials/ Machinery Hire*
1	Remove engine	2	2	£100 per day
2	Rebuild engine	15	1	Total of £9,000 spread evenly
3	Refit engine	3	2	£100 per day
4	Repair dents	5	2	£50 per day
5	Respray bodywork	5	2	Total of £2.5 K spread evenly
6	Fit new chrome	1	1	£1,750
7	Fit wiring loom	3	1	£1200 on day 1
8	Test electrics	1	1	£0
9	Replace bulbs	1	1	£250
10	Take mot	1	3	£525 (incl transport)

Notes:
[a] Labour is charged at £200 per unit per day.

total £100 per day. The remainder of the estimated project costs relating to each task are shown in Table 10.1.

The costs can be totalled up on a daily basis as shown in Table 10.2.

The costs, not forgetting the additional £100 per day project management charge, can now be loaded onto the Gantt Chart as shown in Figure 10.4 below.

The Gantt Chart in Figure 10.4 shows the outflow on each day of the project – the 'Daily Total (£)'. These costs are totalled on the right-hand side and show an overall cost of £28,950. The renovation project therefore, as budgeted, only makes a profit if it can charge the client in excess of £28,950 for the work carried out. If the client has agreed to pay the garage £33,600 upon completion for the restoration, then the project would appear to have a profit margin equivalent to £4,650 or 16%. The one cost though that has not been included, is the cost of financing. If costs must be paid on a daily basis, then they must be funded each day until the client pays the garage. If the cost of finance is 1% per day, charged on the previous day's closing balance whether it is positive or negative, then this will impact on the project's profitability as shown in Figure 10.5 (In reality the interest rate would be significantly lower, but 1%/day is used for illustration purposes).

The cumulative costs, including interest, are now £32,035 instead of £28,950, and the profit has fallen to £1,565 (£33,600–£32,035) or 4.9%.

If payment terms can be renegotiated, not in terms of the amount payable, but simply in terms of the timings, then the finance costs could be reduced. If the client was willing to pay the £33,600 but had agreed a payment profile of a third of this amount at the end of day 7, a third at the end of day 14 and a third at the end of day 21, then the profitability for the garage would increase. The daily cashflows, including the finance charges, can be calculated using the approach shown in Figure 10.6 below.

The impact of the change in payment terms and the setting up of cashflows to reflect Figure 10.6 is shown below in Figure 10.7.

The overall interest charge has reduced significantly from £3085 to £603 as a result of the project being overdrawn less following the renegotiation of the payment terms. The impact of

Table 10.2 Total Estimated Project Costs

Task No.	*Task Description*	*Task Duration (Days)*	*Labour Units per Day**	*Materials/ Machinery Hire*	*Costs per Day (£)*		*Total Costs per Day*
					Labour	*Materials*	
1	Remove engine	2	2	£100 per day	400	100	500
2	Rebuild engine	15	1	Total of £9,000 spread evenly	200	600	800
3	Refit engine	3	2	£100 per day	400	100	500
4	Repair dents	5	2	£50 per day	400	50	450
5	Respray bodywork	5	2	Total of £2.5 K spread evenly	400	500	900
6	Fit new chrome	1	1	£1,750	200	1750	1950
7	Fit wiring loom	3	1	£1200 on day 1	200	Day 1: 1200	Day 1: 1400 Day 2&3: 200
8	Test electrics	1	1	£0	200	0	200
9	Replace bulbs	1	1	£250	200	250	450
10	Take mot	1	3	£525 (incl transport)	600	600	1200

Notes
[a] Labour is charged at £200 per unit per day.

the reduced interest charge, and hence of the renegotiated payment terms is to increase the profit from £1,565 (£33,600–£32,035) to £4,047 (£33,600–£29,553) or from 4.9% to 13.6%.

The projects funding requirements shown in Figure 10.7 are significantly less than those in Figure 10.5 and so the project in its latest guise is more attractive and more readily funded.

The renegotiated payment terms shown in Figure 10.7 show how important a consideration this is when compared with the original terms linked to Figure 10.5. What the project manager has been able to do in Figure 10.7 is to protect the project's cashflow and thereby increase the project's profitability.

The payment terms above are simply related to time. Payments can of course be triggered by events other than the passage of time. Table 10.3 below shows a variety of possible project payment types.

All of the payment terms above will need to be considered when constructing the budget and may well be impacted by industry norms, project contracts or even the type of supplier contracted with. Projects tend to avoid advance payments wherever possible, but if, for example, you have contracted with a small organisation for a costly project component it may be in the project's interests to enter into advance payments in order to protect the smaller entities cashflow and help to guarantee delivery. One way that this could be done relatively safely is to provide payment for the raw materials that the contracted-out task requires but to retain legal ownership of these materials whilst at the contractor's workshop.

Penalty payments can be prohibited in some forms of contract, whilst, in the same contract incentive payments are legal. The difference between the two can sometimes be down to the

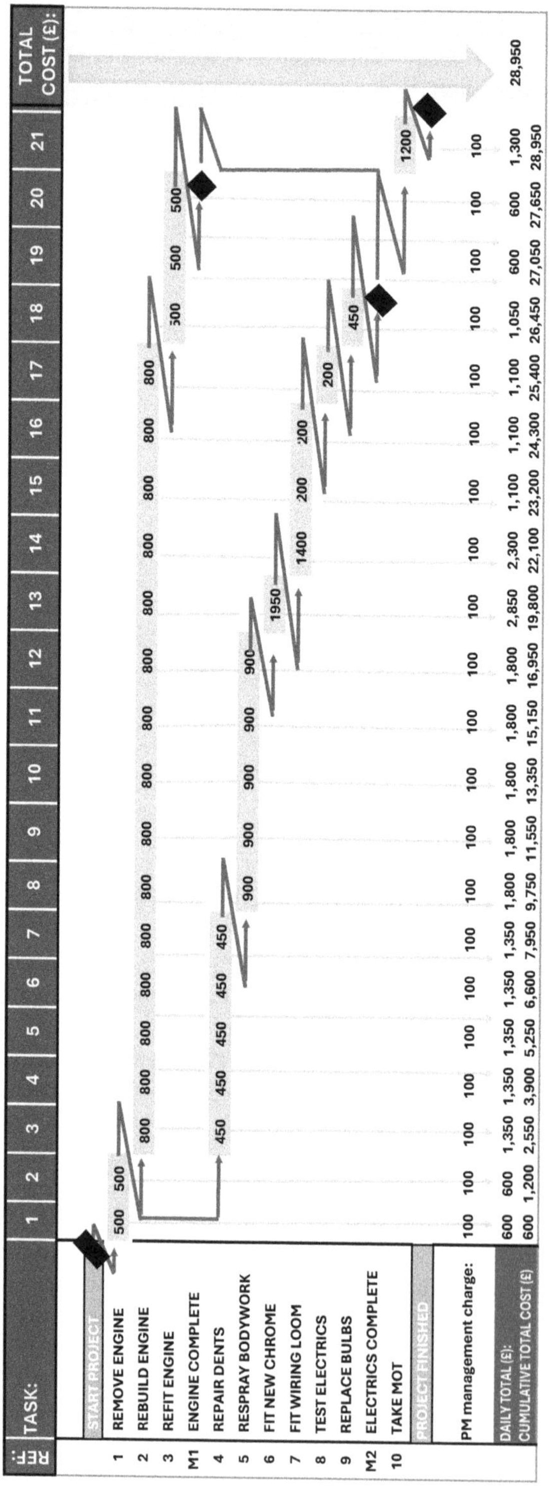

Figure 10.4 Gantt chart with costs loaded.

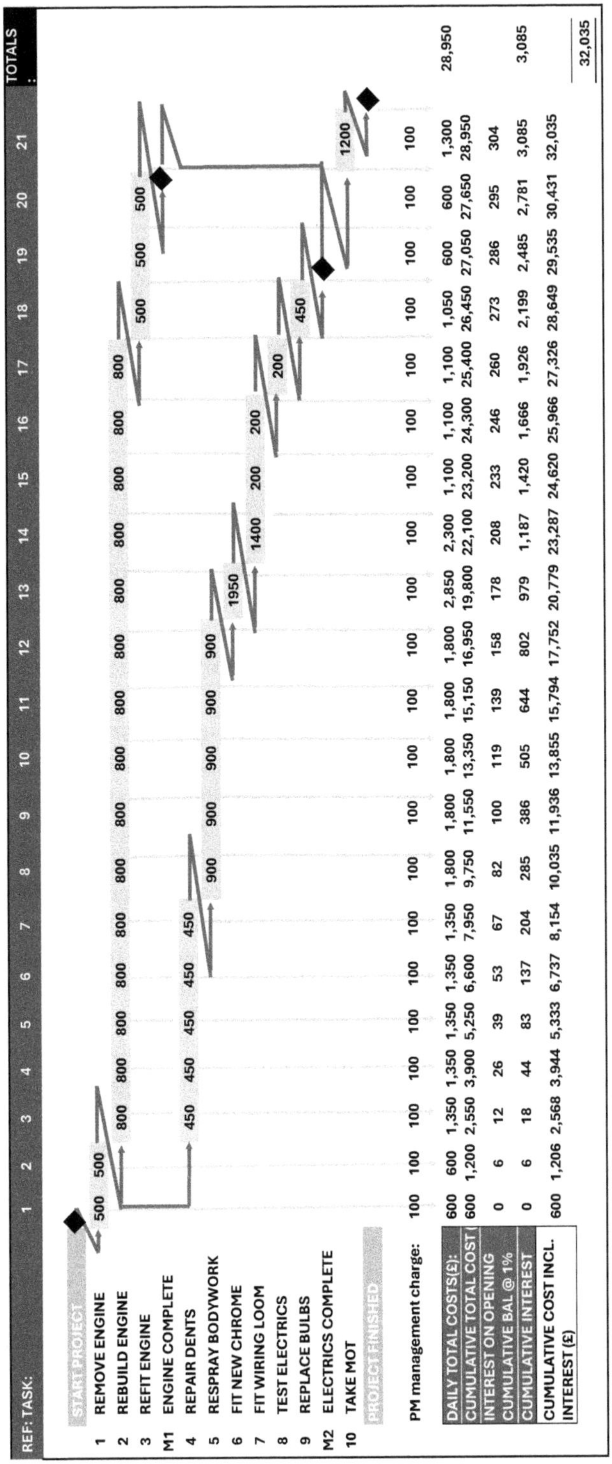

Figure 10.5 Gantt chart with costs and interest payments loaded.

	Starting Bank Balance	
-	Cash paid out	Cashflow in that period
+	Cash receipts	
-	Interest and bank charges paid	
+	Interest recieved	
=	Closing Bank Balance	

Figure 10.6 Setting up project cashflows.

focus or phrasing of the term. Progress payments and milestone payments too are similar. Milestone payments though are more directly tied to the project schedule and would become due once it has been demonstrated that a milestone has been met.

10.2.2 Managing the Cashflow

Having established the project's budget and baselined it, the cashflows projected within the budget need to be managed. The project manager has a number of considerations to weigh up when deciding on the level of sophistication required by the project in terms of cost management. The level of granularity in a cost management process is important and will vary depending on the audience. A cost management model has a number of different levels of activity and focus as shown in Figure 10.8. Each stage is of some value and even stage one on its own offers some benefits but the level of sophistication and of control which the model offers the project increases as each additional level is added.

The first stage is Data Collection. Here the different types of cost need to be considered and data needs to be collected from the appropriate determinants of project costs. These are likely to include fixed costs, variable costs, costs internal and external to the project and both direct and indirect project costs. This represents the most basic form of cost control and allows collection and comparison of basic spend totals against budget totals in a simplistic way.

Stage 2, Cost Accounting allows for costs to be analysed in greater detail addressing where costs have been incurred in terms of the work packages identified in the work breakdown structure or WBS (see Chapter 8), and to set up Cost Accounts by combining the WBS with a similar breakdown structure called the OBS or Organisation breakdown. The OBS details all parties carrying out work on the project both internal and external. The diagram of these Cost Accounts would look similar to the grid at Figure 10.9.

The sum of all the individual Cost Accounts at the planning stage will equate to the project budget.

The benefit of this type of analysis is that by aligning the project's work packages with the bodies delivering the tasks, it can highlight specific areas of project overspend and so enables the third stage of the cost management model – Cost Control.

Stage 3, Cost Control, involves determining, explaining, and correcting cost variances. The data provided by the first two stages outlined above, will provide the raw data for analysis which should be used as the basis for a system of constant monitoring and correction of the project finances. The cost management system needs to be aligned with the organisation's strategic business objectives in order to provide accurate information at the right level of detail in a timely manner. This involves capturing costs early so that any deviation is relatively small and hence any correction should not be too challenging.

REF:	TASK:	1	2	3	4	5	6	7	8	9	10	11	12	13	14	15	16	17	18	19	20	21	TOTALS:
	START PROJECT																						
1	REMOVE ENGINE	500	500																				
2	REBUILD ENGINE			800	800	800	800	800	800	800	800	800	800	800	800	800	800	800					
3	REFIT ENGINE																		500	500	500		
M1	ENGINE COMPLETE																						
4	REPAIR DENTS			450	450	450	450	450															
5	RESPRAY BODYWORK								900	900	900	900	900										
6	FIT NEW CHROME													1950									
7	FIT WIRING LOOM														1400	200	200						
8	TEST ELECTRICS																	200					
9	REPLACE BULBS																		450				
M2	ELECTRICS COMPLETE																						
10	TAKE MOT																					1200	
	PROJECT FINISHED																						
	PM management charge:	100	100	100	100	100	100	100	100	100	100	100	100	100	100	100	100	100	100	100	100	100	
	DAILY TOTAL COSTS(£):	600	600	1,350	1,350	1,350	1,350	1,350	1,800	1,800	1,800	1,800	1,800	2,850	2,300	1,100	1,100	1,100	1,050	600	600	1,300	28,950
	CUMULATIVE TOTAL COST (£)	600	1,200	2,550	3,900	5,250	6,600	7,950	9,750	11,550	13,350	15,150	16,950	19,800	22,100	23,200	24,300	25,400	26,450	27,050	27,650	28,950	
	INCOME FROM CLIENT							-11,200							-11,200							-11,200	-33,600
	INTEREST ON OPENING CUMULATIVE BAL @ 1%	0	6	12	26	39	53	67	-30	-13	5	23	41	60	89	1	12	23	34	45	51	58	603
	CUMULATIVE INTEREST	0	6	18	44	83	137	204	173	161	166	189	230	290	379	380	392	415	449	494	545	603	
	CUMULATIVE BALANCE (£)	600	1,206	2,568	3,944	5,333	6,737	-3,046	-1,277	511	2,316	4,139	5,980	8,890	79	1,180	2,292	3,415	4,499	5,144	5,795	-4,047	
	PROFIT:																						4,047

NB:IN THIS CHART COSTS ARE SHOWN AS +VE AND INCOME IS SHOWN AS -VE

Figure 10.7 Gantt chart with costs, interest and phased client payments loaded.

Table 10.3 Examples of Different Payment Types

Advance Payment	An advance payment is one made ahead of its normal scheduled delivery such as paying for an item before it is received. Advance payments are sometimes required by sellers as protection against nonpayment, or to cover the seller's out-of-pocket costs for supplying the service or product. In projects they can help fund the production of a large component part
Progress Payment	A progress payment approach links payments to the achievement of steps or phases made towards the completion of the overall project
Milestone Payment	Milestone payments are not too dissimilar to progress payments but the trigger point for payment is linked specifically to the completion of predefined project milestones as would be shown in the project schedule
Payment on Completion	Payment on completion is triggered once the project or task linked to the payment has been completed
Retention Payment	A retention payment is the final payment representing the amount held back as a security for performance. It may allow for a period of snagging for a short time after the project is set live
Incentive Payment	Incentive payments are due once the project has reached certain predetermined conditions or levels of functionality. They are used as a means to encourage project performance that hits certain key, stretching, goals
Penalty Payment	A penalty payment is a financial charge imposed on a project for failing to fulfil certain terms of a contract, or agreement. Often employed as a deterrent, these penalties can be used to enforce compliance or encourage stipulated conditions to be met

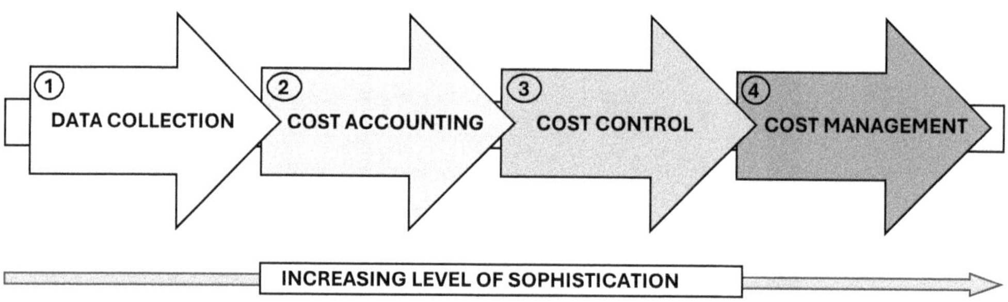

Figure 10.8 A cost management model.

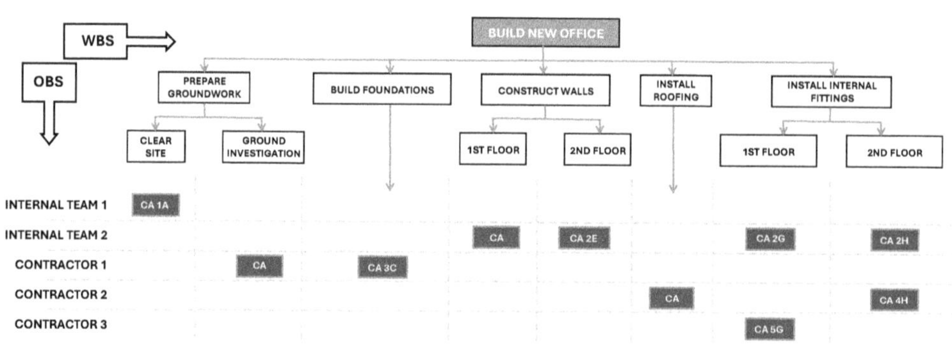

Figure 10.9 Cost accounts set up at the intersection of the WBS and OBS.

Stage 4, the cost management of the projects cashflow, takes the output or conclusions of stages 1-3 and links this to project controls, monitoring such things as actual costs, project progress through milestones and considering the level of costs necessary to complete the project. Earned Value Management, the subject of Chapter 11, is worth reviewing in connection with this.

10.3 Project Review: The impact of the Budget on the Project: The Canadian Firearms project

10.3.1 The Project Background

The Canadian Firearms Registration project is one that may lay claim to have one of the worst project budget overruns in history. This claim is made on Calleam's excellent 'Catalogue of Catastrophe' website[4] which states that the project's original budget of $119 m rose to a cost of £1 bn before the project scope was cut back.

The aim of the project appeared quite simple – as the project name implies, the project focused on the establishment of a register of firearms. The Canadian Government's need for this came immediately after the 1995 introduction of universal firearm registration (the Firearms Act (Bill C-68)), requiring that all gun owners be registered by January 1st 2001, and that all firearms be registered by January 1st 2003.[5] The new laws were built upon what some already considered to be strict regulations which had required handguns to be registered since 1934 and in 1977, required all firearm purchases to be subject to police scrutiny whilst also prohibiting a range of firearms both in 1977 and again in 1991.[5]

Estimates by the federal government put the number of gun owners at 2.46 m and the number of firearms that would need registering at 7.9 m.[6]

In total, the act and associated amendments to the Canadian Criminal Code provided for:

- licensing all firearms owners;
- registering and tracking all firearms and transfers of firearms;
- controlling the import and export of firearms;
- strengthening the offence and sentencing provisions of the Criminal Code, including mandatory minimum sentences for certain firearms offences; and
- prohibiting additional specific firearms.

Supporters of the legislation saw it as a way of increasing public safety through increasing levels of responsibility amongst gun owners as well as helping police in tracking stolen or lost firearms. The counter view perceived it to be a measure that curtailed individual freedoms and that would be largely unworkable and bureaucratic.

10.3.2 The Project Issue

The project cost estimates made at the start of the project were thought in some quarters to be both doubtful and inaccurate. Registration needed cooperation across a number of different federal ministries, for example Customs, Justice, the RCMP (Royal Canadian Mounted Police - Canada's federal police force,) and all three territorial governments.[5]

The 2002 Report of the Auditor General of Canada stated that when reviewing the Costs of Implementing the Canadian Firearms Programme, The Canadian Department of Justice failed to provide Parliament with sufficient information to allow it to effectively scrutinise the project costs and to ensure accountability. Specifically, it failed to provide sufficient financial information and explanations for the dramatic increase in the cost of the project which were deemed necessary. The report went as far as saying that the information provided did not fairly present the cost of the project to the government.[6]

Many reasons were offered by the Department of Justice to explain the cost escalation, 'including major delays in making regulations, provinces opting out of the programme, the need for additional initiatives, incorrect assumptions about the rate that applications for licences and registrations would be received, and an excessive focus on regulation and enforcing controls.'[6]

Assuming the reasons above to be correct, then it means one of two things. Either the budget setting was at fault due to e.g. the 'additional initiatives' having been omitted, or the costs of these elements were severely underestimated. Or both.

Observations from the Auditor General's 2002 Report relating to the control of the budget were as follows:

- In 1996 the Department of Justice established a division called the Canadian Firearms Centre. This was designated as the "single point of responsibility and accountability to implement the Canadian Firearms Programme." It was seen as being critical to achieving delivery within budget.
- The single point of responsibility and accountability concept was not implemented. The Centre only reported expenditures that it incured or reimbursed, and not the full costs of the project.
- The Department of Justice did not provide enough information to Parliament on the initial costs, nor on the escalating costs, to allow it to sufficiently scrutinise the Programme.
- As a high-level initiative, the Canadian Fire Arms project was designated as a 'Major Crown Project'. Under this categorisation, Treasury Board policies require departments, at a minimum, to annually report the following types of information to Parliament:
  - description of the programme,
  - leading and participating departments and agencies,
  - total expenditures to date and planned expenditures for future years to the completion of the project,
  - prime and major sub-contractors,
  - major milestones, and
  - progress reports and explanations of variances.

The Department of Justice as at the time of writing the report had not fully reported this information.

Complete and accurate financial information and explanations on costs and revenues should have been reported to Parliament annually in the Department's Report on Plans and Priorities and in its departmental performance report. These formal documents are intended to provide information to Parliament in order to allow it to scrutinise departmental expenditures. This was not done fully.

Without providing Parliament annually with complete and accurate information on all past, current, and forecasted expenditures and revenues relating to the project and explaining any major changes, the project would appear to have avoided the safety net of budgetary scrutiny.[6]

A further indication that budget and cashflows were not being adequately controlled is provided in the report[6] which stated that 'in our view, the financial information provided for audit by the Department does not fairly present the cost to the government.' This failing was severe enough for the Auditor General's report to state that the audit of this information was stopped 'because we did not believe that a detailed audit would result in substantially different findings'.

The report reveals that one potential issue in keeping to the original budget was the discovery that initial estimates of funds required and expected revenues were based on a series of unrealistic assumptions, and that funding for several of the initiatives required was lacking. Alongside this, in 2001, the Department told the Government that the three-year-old Canadian Firearms Registration IT System was not working well; its technology was expensive, inflexible, out-of-date, and could not be modified at a reasonable costs to support future operations. This is not an uncommon occurrence in projects with a developing design – not only are they difficult to accurately estimate a budget, and then to follow a budget for, but they can end up spending what budget they have on nugatory work as the requirement crystallises.

10.3.3 How Could a Developed Budget and Cashflow Process Have Helped?

Section 10.2.1 above begins by discussing the importance of the WBS in constructing the project budget. This breaks the project down to individual tasks at a level which is granular enough for costs to be monitored and controlled against.

The WBS forms the basis of the projects scope, and this would seem to have been less than fully defined in the case of the Canadian Firearms project. This would have led to a lack of clarity around the baselines of the project, from the perspective of both cost and time. Without clear agreement around a baselined budget monitoring and controlling the project spend becomes difficult. If a concrete budget had been set (initially it seemed as if it had, but the degree of financial escalation would suggest otherwise), then more formal approval would have been required for any changes. The requirement noted above for a single point of responsibility and accountability to implement the Canadian Firearms Programme, was indeed critical to achieving delivery within budget, and the lack of this being created and adhered to culminated in the doubts raised by the Auditor General that the information provided to the project audit did not fairly represent the costs.

The difficulties caused by this lack of certainty around the scope and costs would have impacted the project's ability to follow Treasury Board's approval processes and this in turn added to the project's problems. It is further stated[6] that the approval process for the project by the Treasury Board was aimed at minimising the risk of failure and overspending. The process included the use of a framework allowing comparison of the actual expenditures to those authorised by the Government, which would have managed the cashflow better by, as a minimum, highlighting project deviations from the base line scope and cost at an earlier stage. This would have ensured that a budget was in place and that cashflow management could be aligned with it. Critically the report[6] reveals that 'the Department began implementing the Programme in December 1995, more than two years before receiving either Preliminary or Effective Project Approval'. It is no surprise to read later in the report that there 'were deficiencies in the management of revenues' in particular that information supplied by the project to the audit was not timely and was not complete.

The way in which a systematic budgeting and cashflow management approach would have helped the project is summed up in the conclusion of the Auditor General's Report which recommended the following:

- The Department of Justice, should use a meaningful activity-based reporting framework that should annually provide Parliament, in its departmental performance report complete, accurate, and up-to-date financial and management information on the following:
  - the full costs to develop, implement and enforce the Canadian Firearms Programme,
  - all revenues collected and refunds made,
  - forecast costs and revenues to the point at which the Department expects the programme to become fully operational, including details on outsourcing major components of the Canadian Firearms Registration System and moving certain headquarters functions to Edmonton, and
  - complete explanations for changes in costs and revenues, and changes to the overall programme.

A well-constructed budget and cashflow management system would be able to provide the answers to the above points.

10.3.4 Conclusions on the Canadian Firearms Project

The lack of project definition was found to have directly contributed to costs. In a follow up to the 2002 report the Auditor General produced a second report in 2006.[7] This report stated that the costs of the Canadian Firearms Information System (CFIS) were more than twice its original estimate due to poorly defined initial requirements and poorly managed change control process (see Chapter 12). The impact of these two issues could have been much reduced by a well-structured budget and cashflow management processes. The report gives the impression that both a well-structured budget and cashflow management processes were very much missing, stating for example that 'estimates of final costs varied over time, as well as what was included in calculating them.'[7]

Without a reliable method of keeping track of project costs enabling accurate forecasting of project out-turns, any project will be difficult to manage, and its final outcome will always be at risk without a financial compass.

10.4 Budgeting and Cashflow Management Before the Birth of Modern Project Management*

(* In this section, the research relating to the construction of HMS Barham 1912–1915 mentioned in the introduction is drawn upon to help illustrate how the subject of this chapter has developed.)

Before modern project management was codified and established in canons such as the APM's or the PMI's Body of Knowledge, projects were still being delivered and cost management was central to their success. The construction of HMS Barham at the start of the twentieth century illustrates this.

The Naval estimates for each annual shipbuilding programme were discussed in Parliament. It was a long-established constitutional practice for the Government to request money, the House of Commons to grant it and the House of Lords to agree to the grant. Approval of the estimates gave parliamentary consent, and hence legal authority, to the consumption of resources and spending of money in order for each government Department to finance the Department's spending for the financial year.[8] The estimates once approved enabled a formal budget to be put in place.

Photo 10.1 below shows how elements of the ships build were split into a series of instalments payable to the ship builder by the Admiralty. These were linked back to the contractual terms agreed and show control and planning of cash receipts representing one side of the cashflow.

On the cost side, Photo 10.2 shows how the number of tradesmen were collected on a daily basis (in this case on Wednesday 4/11/1914), for on charging to each vessel and hence each contract. The column on the lefthand side shows different trades such as Angle Smiths, Platers, Joiners etc, whilst across the top are the shipyard numbers allocated to each ship, No. 424 as shown in the top line of Photo 10.2 being HMS Barham. The totals about a quarter of the way

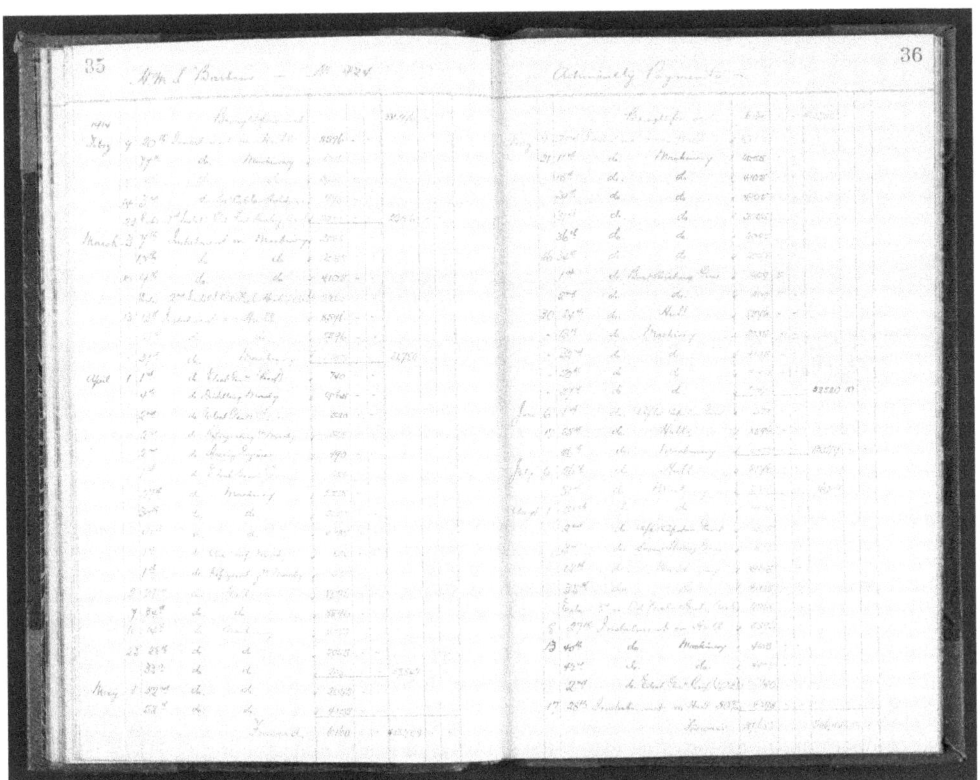

Photo 10.1 Example of the admiralty payment profile for HMS Barham.

Source: By permission of the National Records of Scotland and University of Glasgow Archives & Special Collections, Records of Upper Clyde Shipbuilders Ltd, Clydebank Division collection, GB248 UCS 1/61/2.

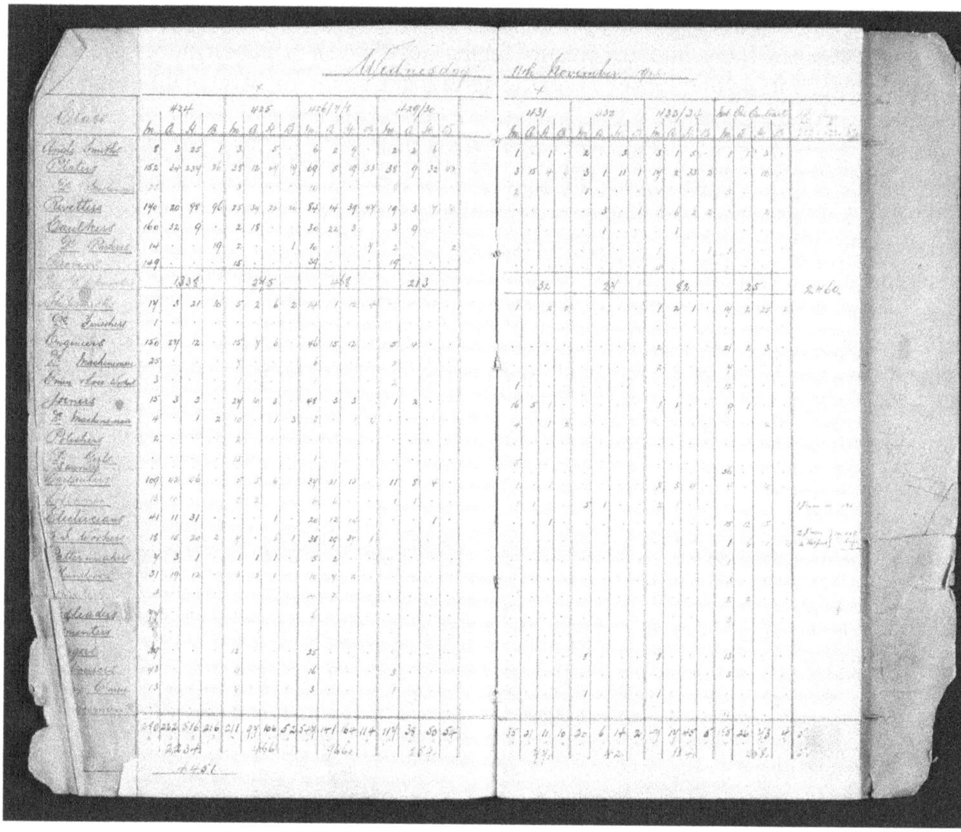

Photo 10.2 Example of the collection of manhour statistics used in the construction of HMS Barham.

Source: By permission of the National Records of Scotland and University of Glasgow Archives & Special Collections, Records of Upper Clyde Shipbuilders Ltd, Clydebank Division collection, GB248 UCS 1/59/1.

down (1,338 in the case of HMS Barham), show the total number of men involved in the construction for the seven trades listed above the total figure.

These types of figures were collected across numerous different types of resources and were compared to the budget estimates with the under or overestimate noted and the total variance noted at the bottom of each page. This was done on a monthly basis across the build of each ship. An example of this for the months of April and May 1915 can be seen below at Photo 10.3.

The comparison of cost against estimates shows that in the construction of HMS Barham, a budget had been set (the estimates) and component costs were collected and aggregated. The comparison of these figures shows that the variances were being noted and totalled to allow the management of the project's cash flow by balancing the income instalments against the project outgoings, which provides direct evidence of monitoring against a set of baselined budget costs.

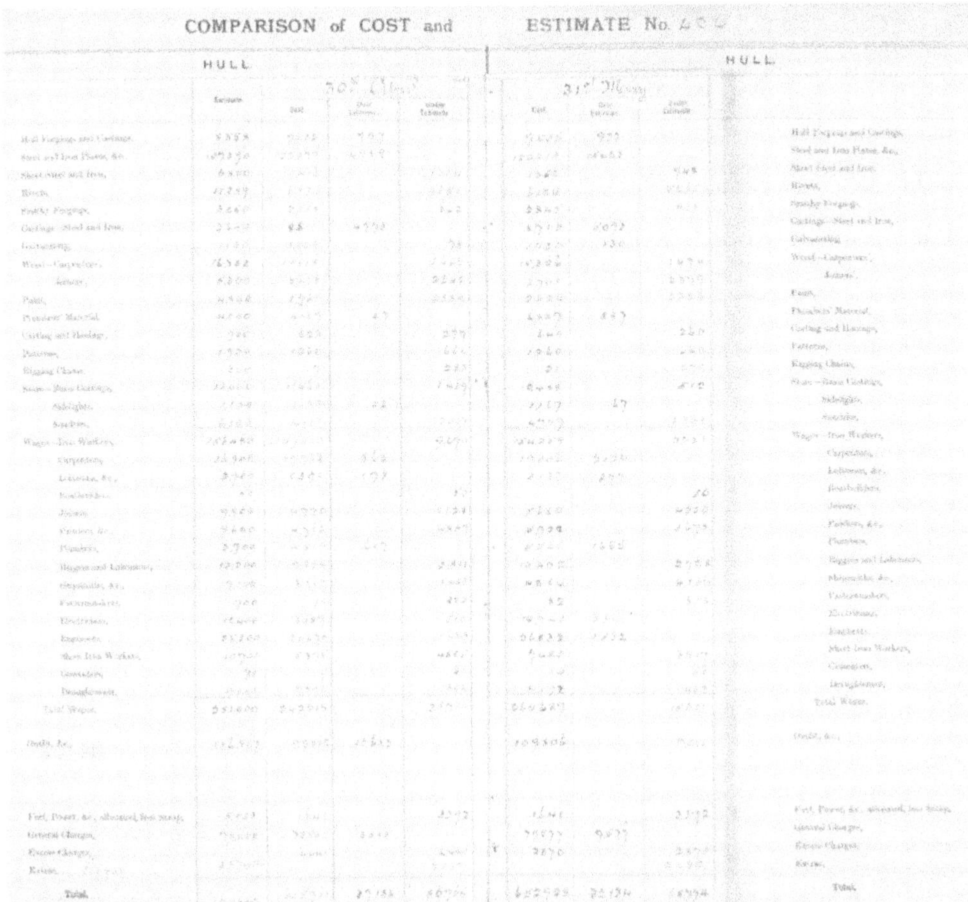

Photo 10.3 Example of the comparison of costs against budget estimates in the construction of HMS Barham.

Source: By permission of the National Records of Scotland and University of Glasgow Archives & Special Collections, Records of Upper Clyde Shipbuilders Ltd, Clydebank Division collection, GB248 UCS 1/86/9.

References

1 APM. (2019). *APM body of knowledge* (7th ed.). Association for Project Management.

2 Project Management Institute. (2021). *The standard for project management and a guide to the project management body of knowledge* (7th ed.). Project Management Institute, Inc.

3 Ohara, S. (2017). *P2M – A guidebook of project & program management for enterprise innovation* (3rd ed.). Project Management Association of Japan (PMAJ).

4 Calleam. (2024). Why do projects fail? *Firearms Registry*. https://calleam.com/WTPF/?p=1949 [Accessed: 19 July 2024].

5 Mauser, G. (2001). Misfire: Firearm registration in Canada. https://www.sfu.ca/~mauser/papers/slipperyslope/FI-Slippery-Slope.pdf [Accessed 12 July 2024].

6 Auditor General of Canada. (2002). *Department of justice—Costs of implementing the canadian firearms program*. Office of the Auditor General of Canada. https://publications.gc.ca/collections/collection_2012/bvg-oag/FA1-2002-2-18-eng.pdf [Accessed 6 July 2024].
7 Auditor General of Canada. (2006, May). *Canadian firearms program*. Office of the Auditor General of Canada. https://publications.gc.ca/collections/Collection/FA3-40-2006-4E.pdf [Accessed 6 July 2024].
8 UK Government. (2023). *HMT main estimates*. https://www.parliament.uk/globalassets/documents/commons-information-office/p06.pdf [Accessed 22 July 2024].

Chapter 11

Progress Monitoring and Control

It is all very well establishing a project plan, agreeing a project schedule and budget, but if there is no project mechanism to monitor and control project progress, the good work done in the project planning will gradually diverge from the actuality of the project to the point that it becomes worthless. Projects must be monitored and deliberate action taken to control them.

11.1 The Purpose of Progress Monitoring and Control

The APM[1] define project control as 'tracking performance against agreed plans and taking the corrective action required to meet defined objectives.' The first half of the definition focuses on the need to monitor a project before considering the control or the corrective part of the process. This process definition also requires a third element, the establishment and agreement of a baseline to monitor actual project data against before deciding on any actions required. The purpose of monitoring against project baselines is therefore to provide accurate project information enabling the reporting of progress and to help ensure that informed decisions relating to the project can be made. The PMI[2] define monitoring as a management rather than a leadership task (see Chapter 6) and recognise that projects need to adapt to their environment as they are delivered, meaning that new tasks may be required or new processes enacted. The introduction of new tasks needs to be controlled as if the number of new tasks is never ending, then so will the project be. Monitoring and Control therefore has an impact on a number of different areas of project management including the change control process (see Chapter 12). The PMI give the definition of Monitoring and Control as the 'processes required to track, review, and regulate the progress and performance of the project; identify any areas in which changes to the plan are required; and initiate the corresponding changes'[2] As a definition it is, arguably, more outward looking than the APM definition since it suggests a more proactive approach involving the search, or identification, of areas that may need changes rather than taking a solely reactionary stance.

P2M[3] Shares the PMI's view of monitoring stating that it involves the collection of information about both internal and external environmental changes surrounding the project as well as information relating to the progress of individual project tasks and milestones (see Chapter 8).

P2M describes the control element as the comprehensive examination of project data to date in order to predict project progress up to the time of project completion, planning necessary measures and ensuring that these are carried out.

From the above descriptions it can be seen that effective project control relies on accurate project monitoring and that project monitoring is pointless without project control. The two aspects are therefore closely linked.

DOI: 10.1201/9781003470748-12

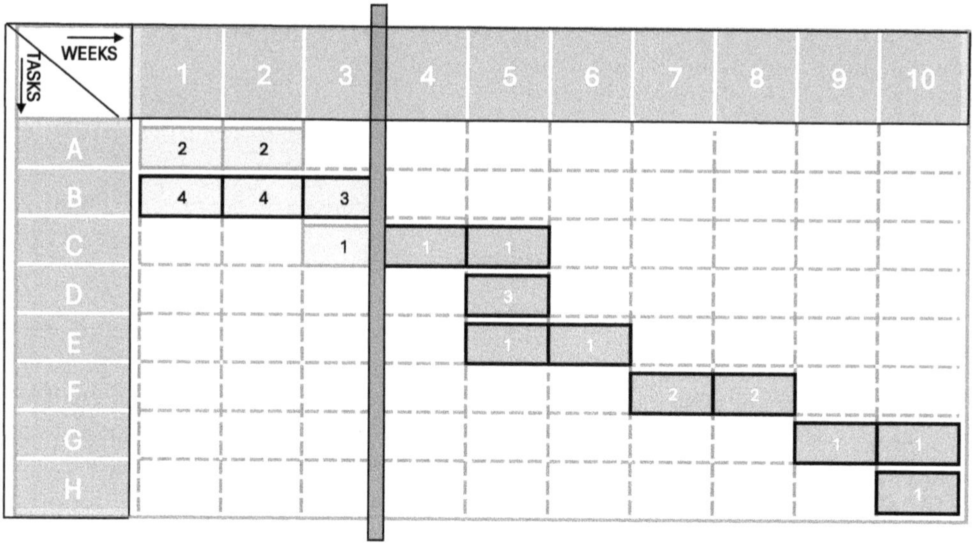

Figure 11.1 An example of a simplified project Gantt Chart.

11.2 How Is Progress Monitored and Controlled?

Project progress is often measured by referring to the project's Gantt Chart similar to the simplified example shown in Figure 11.1.

The chart shows a ten-week project consisting of eight tasks. The dark vertical line represents today's date and all task/weeks to the left of the line have been successfully completed whilst those to the right of it have yet to start. Resource requirements for each week of each task is indicated by the number in each box.

Given this chart, how much of the project has been completed in percentage terms?

This figure becomes important if project reports are being filed and actions are to be based on the answer provided.

What percentage figure was calculated?

Two tasks out of eight are fully completed, so perhaps the answer is a figure of 20%?

The project is three weeks into its ten-week duration, so maybe the answer is a figure of 30%?

Six task/weeks out of sixteen task/weeks have been completed so perhaps an answer of 6/16 or 37% is potentially correct?

If these task weeks were loaded with the resource consumed, then perhaps a figure of 16/30 or 53% is the correct answer?

All four figures ranging from 20% to 53% can be justified, which means that any decision based upon any one of these figures could be questionable and may even be disastrous. Supposing for a different project, the Gantt Chart is supported with comments such as 'we have completed half the project but in just quarter of the time' or 'our actual spend to date is less than our planned spend'. These comments both seem encouraging, but what if in the case of the first statement the outputs are not of the required quality? What if the reason that actual spend is less than the planned spend is because the project has been extremely slow starting and very little of the work has actually begun? What if the project reports that it is on schedule to complete on time but neglects to say that it is 150% over budget?

Each of the statements given above fails to take account of at least one of the three components of project progress, the time taken, the cost incurred, and the quality delivered. These three measures of progress are like a three-legged stool. Any attempt to measure delivery without reporting on all three components will often end in disaster.

11.2.1 Progress measures and Earned Value Management (EVM)

In order to be of use therefore, any report of project progress must reflect the time the project has used up, the budget that the project has consumed and the quality of output that the project has delivered to date. Measurement of these is done in comparison to the baselined figures agreed by the project manager with the project sponsor for each of the three components. Costs will have been baselined in the form of the project budget agreed as part of the business case. Time will have been baselined in the form of the project schedule again agreed with the project sponsor as part of the business case (see Chapter 2), whilst quality will have been defined and baselined as part of the discussion around intended project benefits and the project deliverables that will provide these benefits. These baselines for time, cost and quality form the basis of the project contract, and need to be reported upon using meaningful measures which are fit for purpose.

The comments at the end of section 11.2, reporting project progress, were clearly not fit for purpose as each individual statement did not address the triumvirate of time, cost and quality.

One method of supplying meaningful figures is through the use of Earned Value Management (EVM). The usefulness of this technique has been recognised by the profession and is mandated on certain UK Government project contracts. It has been shown to have numerous benefits, limited not just to reporting. Table 11.1 lists some of the main advantages.

The benefits shown in Table 11.1 can loosely be categorised into three different areas, The first three help to improve the cultural bonds within the project. The commercial bond is strengthened as both customer and supplier need to discuss how EVM is implemented and what within the project is considered 'value'. The second three benefits are key to improving project monitoring, whilst the last two help in improving the level of both accuracy and meaningfulness with project reporting. The seventh listed, and odd one out, impacts all three areas.

Given the demonstrable possibility that the use of EVM can help in project control and delivery, the next step is to review how it works.

EVM has had a reputation in some circles for being difficult to calculate, but it is in fact quite straightforward. There are only three things that must be known with respect to project progress:

1 The Planned Value
2 The Actual Value
3 The Earned Value

Table 11.1 Benefits of Using EVM

Benefits of Using Earned Value Management
It reinforces and spreads best practice in project management
When shared with suppliers, it strengthens the commercial bond
It facilitates a structured and robust reporting process
It provides accurate and timely data for management action
It establishes clear links between activity scheduling and costs
It forces project managers to consider and address budget and resource issues
It promotes an integrated approach to project management
It gives visibility of risk contingency and its consumption
It provides the basis for an independent assessment of likely outcome in terms of cost and schedule

This becomes even more straightforward when the definitions of terms 2 and 3 are laid out. The Planned Value comes from the baselined schedule and states how much value the project schedule indicates should have been delivered by a specific point in time. The Actual Cost asks how much has been spent by the project at a specific point in time. The Planned Value can be read straight from the agreed, baselined, schedule and the Actual Cost can be ascertained from the project accounting function. This just leaves the Earned Value to be defined. A useful definition of this is:

Earned Value = the value of the useful work done, at any point in the project, expressed in terms of the budget assigned to that work.

In other words, how much work has been completed that contributes directly to the project benefits being delivered, and what was the cost assigned to this work in the baselined budget?

Once the terms are defined, the relevant values can be calculated and then all that EVM requires is that the relationships between the terms are investigated.

11.2.2 How EVM Works

Suppose, in an example of a very simple project, Project A, a thousand bricks are to be laid at the rate of 100 per day and that the total cost for this is £1,000, or £1 per brick. If after five days work, a bill is received for all work done to date of £350 is this good news for the project?

What the project manager knows to date about this project is the Planned Value after five days. The Planned Value was a £1 per brick laid at a rate of 100 bricks a day, so the Planned Value at day five equals £500 (Figure 11.2):

And from talking to the project accountants, the project manager knows that the Actual Cost to date is £350 (Figure 11.3):

The remaining EVM variable is the projects Earned Value. Using the definition of earned value 'the value of the useful work done, at any point in the project, expressed in terms of the budget assigned to that work', the amount of earned value actually delivered by the project can be calculated. How is this done for the project example above? If the number of bricks correctly laid is counted and priced at £1 for each brick, then this will give a figure for the value earnt by the project up to this point. Assume this was done and it was found that only 250 bricks had been laid, then the

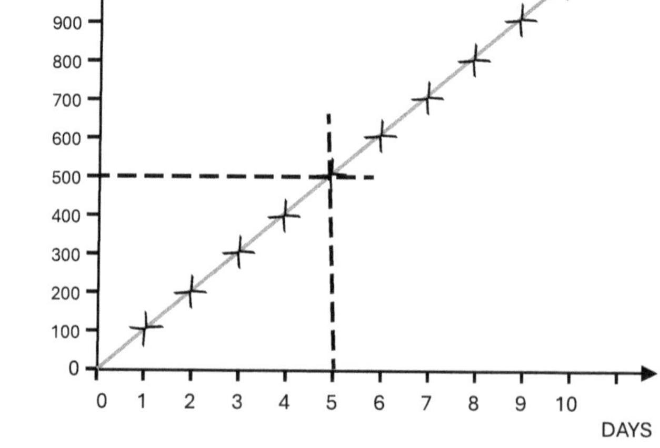

Figure 11.2 Project A: planned value after 5 days.

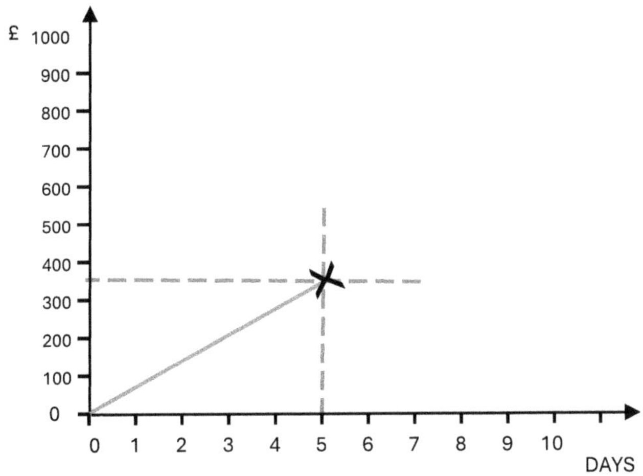

Figure 11.3 Project A: actual cost after 5 day.

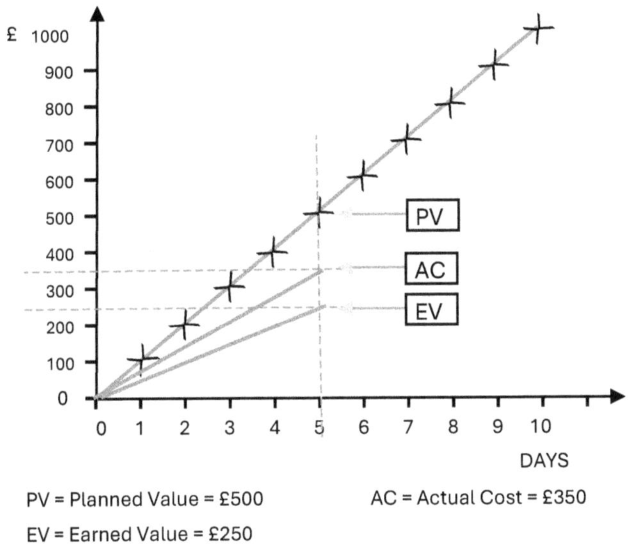

Figure 11.4 Project A: planned value, actual cost and earned value after five days.

Earned Value for the project to date would be £250. If this figure is plotted along with the Planned Value and the Actual Cost, then the values would be as shown in Figure 11.4.

The value shown in the graph for Earned Value (EV) is less than the amount of work that had been planned to be completed by this time (PV) so the project is behind schedule. Similarly, the Actual Cost (AC) of the work done is greater than the value of that work, done expressed in terms of the original budget assigned to it, so given the amount of work done, and the amount billed for this work, the project is over budget.

Having calculated PV, AC and EV the relationships between the value for each of the three terms can be investigated.

The first way in which these relationships need to be analysed is simply by looking at the variances between the terms.

The first of two EVM variances is the Cost Variance (CV) which is calculated as:

$$CV = EV - AC$$

Looking back at the graph in Figure 11.4 shows the values for EV as £250 and for AC, £350.

The variance is therefore:

$$CV = 250 - 350 = -100$$

The variance is less than zero which means as noted above the project is over budget for the amount of work done at the end of day five. If it was on track then of course, EV would equal AC and give a CV of zero. A positive CV means that the project is under budget given the amount of useful work completed.

The second variance that should be considered is the Schedule Variance (SV). This evaluates how well the project is doing in terms of useful work completed compared to the amount of work planned to have been completed by this time – the PV.

The formula is not dissimilar to that for CV in that it starts with EV, but this time compares it with the project's PV. It can be calculated, again using the values given in Figure 11.4 as:

$$SV = EV - PV$$
$$SV = 250 - 500 = -250$$

As per the CV calculation a negative value suggests poor project performance in terms of adherence to schedule, zero would mean the project was on track whilst a positive variance would mean that the project was ahead of schedule. A figure of −£250 shows that rather than delivering the planned amount of useful work, at the end of day five it has only delivered half of what was expected and so is £250 behind the value presented in the planned schedule.

What does this mean for the project manager, what should they do as a result of these figures? One thing that could be done is to plot the values for EV, PV over time. This would potentially produce a graph similar to the EV Chart (from a different project) shown in Figure 11.5

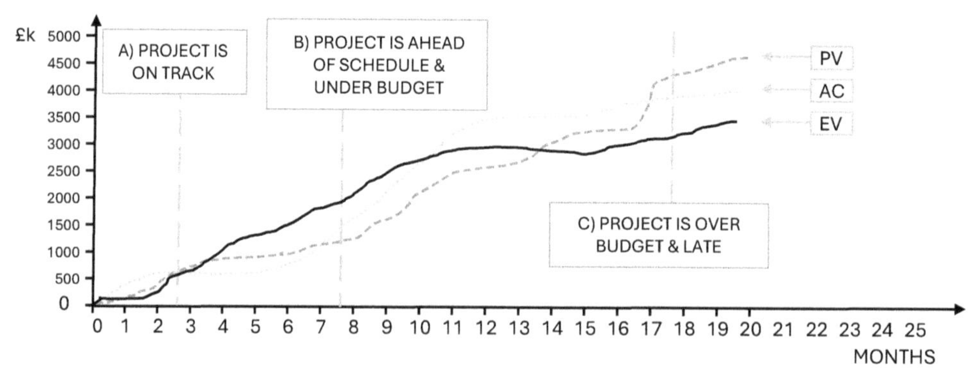

Figure 11.5 An example earned value chart.

If the project manager were to plot the values for the three EVM variables each month, then the resultant graph could show a picture similar to that above. At point A the project is on track, the CV and SV are zero and the project is delivering as planned. At point B the EV is greater than either the PV or the AC which is a healthy position. At point C however, the situation has reversed, and EV is now below the PV and also lower than the AC. This means that the project manager will need to take action if the project is to deliver to time and schedule.

Another combination is possible where EV is greater than AC but less than PV, so the project is ahead of budget, EV > AC, but it is behind schedule, EV > PV. The project manager here has options. They could take some of the savings on the cost side and use them to, for example, increase labour and bring the project back on schedule, or they could let things run. On the face of it, this second option, just letting the project continue behind schedule and under budget, does not seem as if it would be a responsible thing to do. But it would be if the client was far more cost sensitive than time sensitive, or if something had happened within the environment into which the project was delivering which made the project far less time sensitive. This can happen, for instance, when the project is part of a group of related projects and one of the other projects is delivering something that needs to be in place in order for our project to be of use. An example of this could be an aircraft hangar being built to house a new aircraft. If the project building the new aircraft is delayed, then the need for the hangar becomes less urgent.

In order to make the correct decision, the project manager needs to be fully aware of the up-to-date parameters of their project and will need to discuss the options with the project sponsor so as to ensure nothing has changed in terms of the Sponsor's requirements.

Knowing the cost and schedule variances enables the project manager to determine how the project is performing. But it is also very useful in answering the question 'how much better or worse is the project doing than the plan?' This enables a view to be taken on possible project outcomes (which will be revisited below), and it also gives a starting point if assessment of the accuracy and validity of the projects budget or schedule is needed.

EVM offers the potential to judge how well the project is performing against the project's budget and schedule by calculating two important performance indexes.

The Cost Performance Index (CPI) and the Schedule Performance Index (SPI) do exactly this. The formulas are similar to the variance formulas except that rather than having a minus sign on the right-hand side there is a divide sign, meaning that the CPI and SPI are given as ratios. The formulas are:

$$CPI = EV / AC$$

and,

$$SPI = EV / PV$$

These are interesting ratios as for any point in the project they compare, in the case of the CPI, the value earnt as a result of spending the outlay to date. If the two figures are equal, then the project is on track and the CPI will be equal to one. If the project is performing better than planned, then EV will be greater than the AC, and so the CPI will be in excess of one. Conversely a poorly performing project will have a value of less than one. If the values of the SPI were considered then this would tell a similar story, but in respect of the schedule or time spent on the project.

Using the figures for EV, AC and PV given in Figure 11.4, the CPI and SPI can be calculated for the project.

$CPI = EV \mid AC$
$CPI = 250 / 350$
$CPI = 0.71$

$SPI = EV / PV$
$SPI = 250 / 500$
$SPI = 0.50$

As we have already seen in the discussion following Figure 11.4, the project is underperforming, and the CPI and SPI values are consequently both less than one. A CPI of 0.71 tells the project manager that for each £1 spent on the project only 71 pence (£0.71) of value has been delivered. Similarly an SPI of 0.5 shows that for every week spent on the project only half of the work expected is being achieved.

The CPI and SPI are therefore measures of how efficiently the project is using money and time respectively. In the example above, money is being used more efficiently than time, as the CPI figure is closer to a value of one than the SPI figure is.

Just as it was possible to draw an EV Chart, it is possible to plot a CPI/SPI graphic. Suppose that the project CPI and SPI calculations were such that it gave the following graph:

Figure 11.6 shows that the project is initially performing badly and between months 2 and 3 although the CPI improves the SPI deteriorates further. After month 3 performance dramatically improves both in terms of CPI and SPI. The improvement in the CPI is particularly dramatic, so when the CPI falls between month 4 and 5, the project manager could potentially look back to the improvement at month 3 and identify actions taken at that time so that potentially these can be repeated and hopefully have a similar impact on the CPI. The aim of the project manager

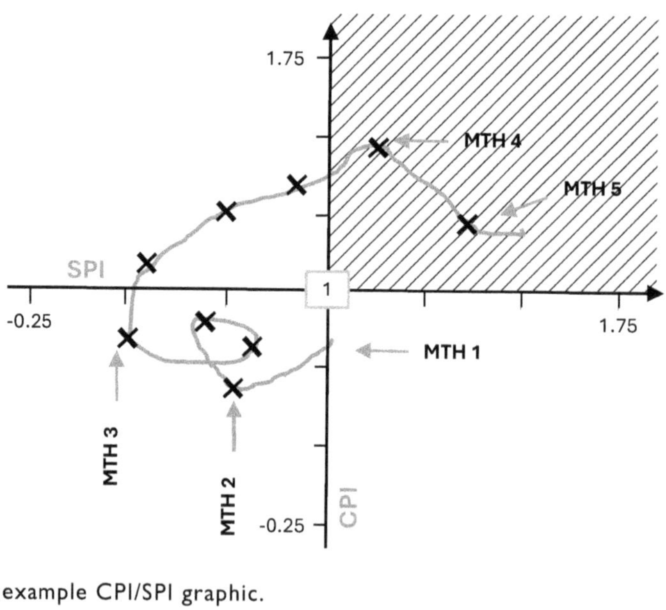

Figure 11.6 An example CPI/SPI graphic.

will be to have the project mapping in the top right-hand corner of the diagram, shown shaded in Figure 11.6, where both the CPI and SPI are performing well with values in excess of one.

One consequence of being able to establish how efficiently either time or money is used within a project, is that predictions of the final out-turn costs and durations can be made. The example given at the start of section 11.2.2 gives figures for:

$$EV = 250$$
$$AC = 350$$
$$PV = 500$$
$$CPI = EV / AC = 0.71$$
$$SPI = EV / PV = 0.5$$

In addition, the budget at completion (BAC), or baselined cost for the project was given as £1,000 and the Time at Completion (TAC), or baselined schedule, was expected to be 10 days.

If the baselined figures are divided by how efficiently money or time are being used (the CPI or SPI), then the following can be calculated:

Estimated Cost at Completion (EAC) = BAC / CPI = £1000 / 0.71 = £1408

and

Estimated Time at Completion (ETC) = TAC / SPI = 10 days / 0.5 = 20 days.

EVM can therefore be used to make out-turn predictions for cost and time.

There is one key assumption made in calculating both EAC and ETC. This is that the CPI and SPI remain constant. If these are both under a value of one for instance, then the project manager will obviously be trying to improve them throughout the project and hence the out-turn predictions of the EAC and ETC may not come to fruition. However, if the values are above one for the CPI and SPI, and they can be maintained at these levels, then the project will come in early and under budget. This could be vital information in planning the start date and scope of a follow-on project. The EAC and ETC values can therefore be seen to provide useful indications and help in project decision making.

The calculations above have taken the project as a whole for illustrative purposes. EVM is though generally calculated at a task level and the EVM statistics for each task need to be aggregated in order to give figures for the project as a whole.

Suppose a project had just three tasks within it. The PV for these three tasks at any given point of the project, could be taken from the baselined schedule and added up to give the overall PV for the project.

Similarly, the individual AC for each task could calculated and then aggregated, as could the EV for each task. This would give statistics for the project's overall AC and EV, which could then be manipulated as in the above examples to give overall project EVM statistics.

11.2.3 Using the EVM Statistics

The collection and calculation of EVM statistics can be managed through a similar process to the illustration of A Cost Management Model shown previously in Figure 10.8. Instead, though

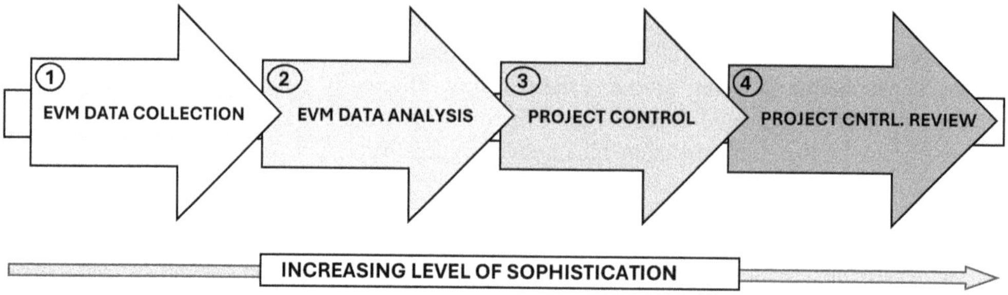

Figure 11.7 A project control model using EVM.

of using costs as part of a Cost Management Model, the EVM statistics can be used as part of a project control model.

When using EVM to help control and deliver a project, the first stage is data collection. This involves the collection of the project information, including establishing baselines, before moving onto the second stage where it is analysed using the calculations described above.

The third stage, project control, involves the application of the EVM Data Analysis to the project, with the result that any poorly performing areas of the project are highlighted and the relevant corrective actions are initiated. The impact of these actions can be recorded and monitored using an EV Chart (see Figure 11.5) or a CPI/SPI Graph (see Figure 11.6).

Stage four of the Project Control Model is the review of the measures taken in the previous stage to ensure they have been successful and to predict future project trends using the information provided by steps 1–3.

Stages one and two could be used in isolation but would only offer the benefits of a reporting function. The addition of stage three facilitates corrective actions driven by the output from stages one and two. Stage four provides a check on the result of actions laid out in stage three and points ahead to out-turn costs and durations.

11.3 Project Review: The Impact of Progress Monitoring and Control on the Project: The Scottish Parliament Building

11.3.1 The Project Background

In 1997, a Government White Paper was published which evaluated the cost of providing the new Scottish Assembly with a "home fit for the Scottish Parliament". The building was to be ready for the 2001–2002 session of Parliament, which began on the 12th May 2001, and the project was estimated at a cost between £24.5 and £34 million.[4]

Since its beginnings, the project has been well documented both through academic papers and official reports.[5] The project's primary aim was to provide a home to house the Scottish Parliament, plans which had been in place for a number of years. Construction of the building began in June 1999 and Members of the Scottish Parliament (MSPs) held their first debate in the new Parliament building on September 7, 2004.

The new building combined the debating chambers with the administrative support offices which had previously been on separate sites. Overall, the new site brought together these different elements into one purpose built parliamentary complex, housing 129 MSPs and in excess of

1,000 staff and civil servants within a 4-acre (1.6 ha) site, east of Edinburgh city centre on the edge of the Old Town. The design of the building, constructed from a mixture of steel, oak, and granite, was hailed as one of the most innovative designs in the country.

11.3.2 The Project Issue

The new Parliament project was intended to be a symbol of devolution and national distinctiveness, and it has broadly achieved this. However, the project suffered significant setbacks due to cost escalations and programme delays.[6] The project was subject to project audits, carried out during the running of the project of which Robert Black, the Auditor General for Scotland, said 'it is unusual to conduct an audit examination of a project that is still running' but, it is 'important that we learn the lessons from this experience in order to ensure that the project is well managed for the remainder of its life.' This indicated that the project, whilst being delivered as of the date of the report, was acknowledged not to be running efficiently, but that instead it was subject to escalating costs. A particular concern was noted that there should have been formal procedures to control changes, based on a detailed cost plan agreed at an early stage, yet again, as of the date of the report (September 2000), a firm cost plan was yet to be agreed. This meant that no project cost baseline had been established so there was no baseline against which to monitor or control rising costs. The shortcomings in the arrangements for cost reporting were a source of project concern expressed both in the Parliament and by the public. This was fully understandable and remarkably there was not even an arrangement in place as part of the construction contract that required project management to provide full cost information to the client on a regular and systematic basis.

In light of these facts, it is unsurprising that the original budget, reported above as being between £24.5 and £34 million, was exceeded, and that instead estimated total costs, as of 2005, were in excess of £450 million,[4] an increase of between 13 and 18 fold. The opening date was also impacted by these issues, with the operational date being delayed by just over three years.

The cause of this can be traced back, at least in part, to a lack of project controls and hence a lack of effective monitoring.

11.3.3 How Could a Developed Progress Monitoring and Control Process Have Helped?

Whilst as with most projects that experience difficulties, it has been stated that 'it is the combination of circumstances that builds up to the final result',[4] the same source attributes much of the cause of the problem to the lack of monitoring and control, saying that there was 'no baselining and that technical plans, budgets, and time scales were a mixture of wishful thinking, political messages and optimistic forecasts – a mixture of budgets so variable and time charts so pliant that they could never be used for control nor for developing corrective actions.' In addition the source also reveals that in 'the interest of speed, many stages were run in parallel, by-passing all controls.' The fact that a project of this size could be set up without 'an arrangement which required project management to provide full cost information to accountable officers or to the client on a regular and systematic basis'[7] is also very surprising. There were various reasons for this[6] but given the high-cost estimates provided by the expert cost consultants, the fact that these weren't reported impacted even more heavily on the client's ability to make informed decisions. The lack of full cost reporting from the outset is acknowledged as having contributed to misunderstandings about the cost of the project and somewhat understatedly it is reported that 'it is not in the best interests of decision making for there to be misunderstanding about the full costs'[7] implying that poor cost reporting results in further compounding of project mistakes, making a bad situation worse. Poor monitoring and control will inevitably mean not that the projects difficulties will go away, but that they will be unearthed at a date later than they could have been meaning that they will again, inevitably, have grown larger and have a more adverse impact on the project's success.

The Scottish Parliament Building project demonstrates the fundamental importance of providing a detailed cost and time baseline for any project. Without them, the size of any change cannot be determined, meaning that the impact of any requested change cannot be measured and the divergence from any agreed budget or schedule cannot be calculated. This lack of ability to quantify an impact means that it is difficult to exclude any requested change on financial grounds as the financial impact on the overall project budget cannot be established or validated. The danger of this, is that the project costs begin to spiral out of control as additional items are included in scope without the full impact being assessed, and any attempt at reporting or controlling the deviations are clouded by uncertainties.

One of the key duties of a project manager is to add clarity and certainty to the project as it is delivered. One of the fundamental ways of doing this is to baseline parameters such as time, cost and quality, at as early a stage as possible. Even if the baseline is not, at the outset, a 100% accurate reflection of the final project deliverables, it can be used through a formal change control process (see Chapter 12) to help direct the project's development rather than allow the project to develop in a potentially organic and unchecked manner. Without the baseline project progress reporting becomes less focused and the correct items are less likely to be reported and hence less likely to be monitored and actioned effectively.

11.3.4 Conclusions on the Scottish Parliament Building Project

There is a wide consensus that the construction of the Scottish Parliament building was a success from an architectural sense and in terms of the building it delivered. Where there is no doubt that it was far from a success is in terms of its overall cost.

Costs were estimated for the project, but the project team did not baseline them or gather and systematically report the total costs to the client on a routine and consistent basis. It is noted[7] that the successful delivery of a project involves the adoption of proven, systematic procedures for progressing the project. These procedures focused on project monitoring and control, are detailed and recommended by the HM Treasury guidance[8] but there were areas where the application of these processes to the Holyrood project was incomplete. More robust application of these approved processes would have significantly reduced the risks faced by the project. Notably, referring back to Figure 11.7, the analysis of the data collected on the project's progress was compromised by the lack of a baseline. It was questionable whether the correct data was being collected in that it didn't enable EVM analysis as a project of this value should. The absence of this capability made it more difficult to highlight those areas of the project which were performing poorly meaning that relevant corrective actions were more difficult to identify and to take.

The lack of an agreed baseline at the start of the project, and 'the routine provision of good quality, high-level information about project costs and performance, and formal change control procedures complying with good practice guidance were not established until June 2000',[7] 3 years after the original Government White Paper highlighted the issue.

Delivery of the project was not straight forward. Some of this was down to the political aspects of the project and some was due to inadequate project management and control. One illustration of the political aspirations impacting the project was the restatement of the project's objective from initially providing “a building suitable for the Scottish Parliament” to “a design worthy of the hopes and aspirations of the Scottish people.”[4] These are very different in terms of implied scope and costs and should have been addressed through formal project management reporting and change control processes. The dangers of running a project in this manner are highlighted in an article that reflects on the project delivery,[4] and asserts that ‘although project control and vision are not incompatible, stakeholders must understand that the more enchanting the vision, the more likely it is to make them forget the rules. This therefore requires strict controls to turn the a vision into a reality and to ensure that the dream that gets out of control does not rapidly become a nightmare.’[4]

11.4 Progress Monitoring and Control Before the Birth of Modern Project Management*

(* In this section, the research relating to the construction of HMS Barham 1912–15 mentioned in the introduction is drawn upon to help illustrate how the subject of this chapter has developed.)

There is much evidence of progress monitoring and control in the shipyards during the time of HMS Barham's construction.

Within the documents reviewed, there were over twenty-five examples of different records detailing various forms of project reporting. These include the production of baseline estimates, reports of shipyard build progress, launch and delivery dates for each vessel, and monthly financial details on ‘contracts in progress’ within the yard. There are examples too, of project control, test reports and follow up records of corrective actions taken, letters between the Admiralty and the yards requesting extra workers and requests for shorter delivery times.

Perhaps though one of the most interesting examples is the monthly totals recorded in the yard's Contracts in Progress book which have been extracted to produce the graph at Figure 11.8 below.

The black lines in Figure 11.8, represent the cumulative costs (or ‘actuals’) spent on the hull, the machinery and the total cost incurred. The lighter grey lines show the original total estimates for the components as per the graph's legend.

The estimates are largely constant and as they don't vary much from inception to competition of the vessel, can be taken as fairly accurate over their lifespan. At the start of the process, the hull was estimated at £634,700 (31st January, 1913) and the estimate varies by 9.5% at the end of the chart (31st March, 1916). The machinery is even closer in terms of its estimate at the start, £410,773 and its final 31st March estimate of £427,504, a difference of 4%. Overall, the combined, or total variance, between the estimate from the beginning of the process to the end is 7.3%, which suggests that there must have been an accurate estimating process in place during the 1912–1916 build process, and relatively few cost surprises.

It is fascinating that all the information needed to calculate Earned Value Management (EVM) statistics is present within Figure 11.8. Actual Costs = Total Costs, Planned Costs = Total Estimates (if scheduled over time,) and earned value = Cash Received, are all shown.

This information in Figure 11.8 comes from a single page of the 1913–1915 Contracts in Progress book. It can't be established if the information was used to monitor and control projects using a method similar to EVM. It is unlikely that it was, as EVM wasn't formalised until many years later and wasn't widely used within UK MoD projects until 2006 when its use was mandated on larger projects. Yet all the information needed to calculate EVM is there, on one sheet, 100 years prior in 1913–1915. The information found shows that monitoring and control was very much present within the construction of HMS Barham, and there is evidence too of corrective actions being taken.

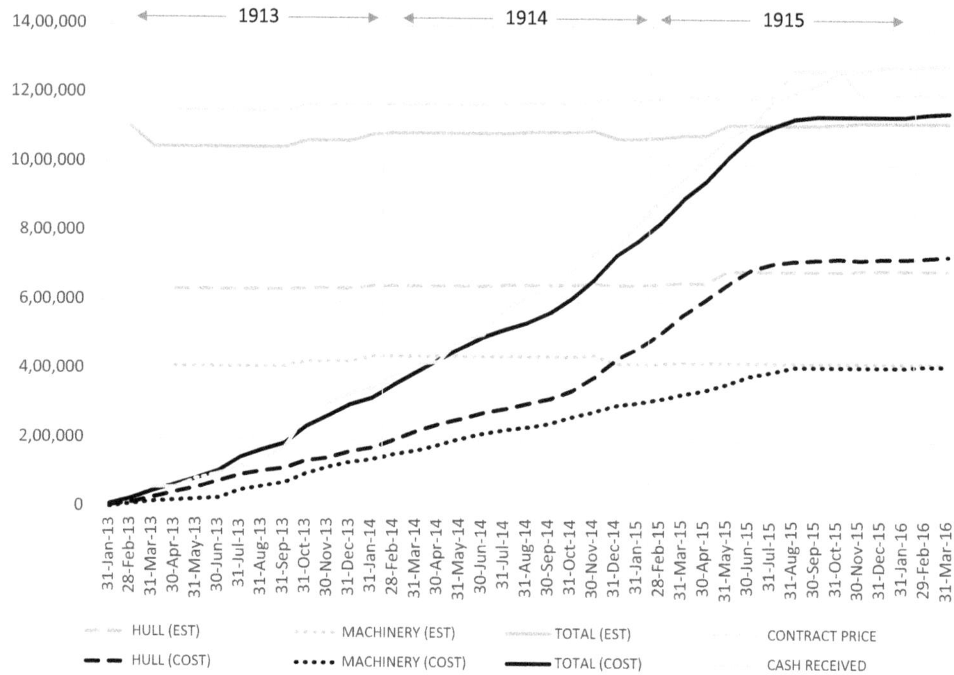

Figure 11.8 Monthly extracts from the contracts in progress book.

References

1 APM. (2019). *APM body of knowledge* (7th ed.). Association for Project Management.
2 Project Management Institute. (2021). *The standard for project management and a guide to the project management body of knowledge* (7th ed.). Project Management Institute, Inc.
3 Ohara, S. (2017). *P2M – A guidebook of project & program management for enterprise innovation* (3rd ed.). Project Management Association of Japan (PMAJ).
4 Piney, C. (2005). Plan has merit—It cost: A project manager reads inquiries into the construction of the Scottish Parliament in Edinburgh. Paper presented at PMI® Global Congress 2005—EMEA. Project Management Institute.
5 Czarnocki, V., & Murray, M. (2004). The new Scottish parliament project: A content analysis of two broadsheet newspapers. https://strathprints.strath.ac.uk/43107/1/ar2004_0401_0408_Czarnocki_and_Murray.pdf
6 The Scottish Parliament. (2000, September 19). Scottish Parliament audit committee Tuesday. https://www.parliament.scot/chamber-and-committees/official-report/search-what-was-said-in-parliament/hscs-19-09-2000?meeting=28&iob=9. [Accessed 22 September 2024].
7 The Auditor General of Scotland. (2000). *The new Scottish Parliament building.* Audit Scotland.
8 UK Government. (1997). *Procurement guidance 2 value for money in construction procurement.* https://www.thenbs.com/PublicationIndex/Documents/Details?Pub=HMT&DocId=247719 [Accessed 22 March 2024].

Chapter 12

Benefit Management, Change Control, Client Satisfaction, and Project Success

The four topics in this chapter may at first seem not to be closely related. However, if benefits are not tracked and managed, then appropriate changes cannot be made to the project delivery, meaning that clients may not be satisfied with the results and hence the project could be deemed a failure. The effectiveness of a project is directly related to the level of benefits delivered by that project, and they should consequently be the focus of delivery.

12.1 The Purpose and Definition of Benefit Management, Change Control, Client Satisfaction and Project Success

The purpose of the benefit management process, as defined by the APM,[1] is to ensure that the project is focused on all stages of the benefit delivery process from the initial identification of the benefits to the planning, monitoring and ultimately the delivery of the benefits themselves. The PMI have a similar focus, describing the Benefit Management Plan as providing the definition of the process for 'creating, maximising, and sustaining the benefits' delivered by a project,[2] a key part of which is the continual assessment of whether project deliverables are on track to deliver the project's planned benefits. This involves not just benefit delivery but benefit management too, ultimately comparing the planned benefit delivery with the actual benefits delivered.

The need for a thorough change control process is laid out clearly in the P2M[3] document. Here it says that a register of change requests for the project as a whole should be maintained. It should evaluate each change request from the standpoint of benefits introduced by the proposed change, balancing them against alterations to the time, cost, quality and risk parameters before obtaining approval prior to implementation. The APM[1] reflect this approach defining change control as a process resulting in requested changes being identified and evaluated before being approved, rejected or deferred.

The PMI[2] make the clear distinction between change control and change management, highlighting that change control is the process of reviewing the impact of a proposed change and thereby recommending whether it should be implemented or not, whereas change management is the overall process of 'transitioning individuals, groups, and organisations from a current state to a future state in which they realise desired benefits.' Change management is therefore the entire process of project delivery, and it is change control which is discussed in this chapter. There will inevitably be changes throughout a project's lifecycle, particularly for those projects with a more incremental, rather than serial, lifecycle (see Figure 4.9). The evaluation of changes helps protect the project against scope creep and helps to ensure that only changes that add to the project outcomes are actioned.

DOI: 10.1201/9781003470748-13

The client, or customer, is defined similarly by the APM,[1] PMI[2] and P2M[3] as the project owner, the entity that commissions and pays for the work, or the individual or group who has requested, or is funding, the project. The client may not always be the one who ultimately pays for the project, although they will have an influence over its delivery. Whether or not the client pays directly for the project outputs, the difference between the two, is, for the purposes of this discussion, hypothetical, and the client can be regarded as both requesting and paying for the project, in addition to using the project outputs. Without some benefit being defined and agreed as desirable by the entity providing the finance, the project would not go ahead, so the two are closely linked.

Customer satisfaction can be difficult to measure as it can take many and varied forms.

From a project viewpoint, customer satisfaction is often evaluated by assessing the effects or by evaluating the value to the customer of the project outputs[3] and hence the customer's level of happiness. The APM[1] in reviewing a project's success criteria, define them as the 'satisfaction of stakeholder needs'. This is a wider definition than directly satisfying just the client's needs, but does infer that through the maximisation of satisfaction relating to stakeholder needs, the client will also derive benefit and in turn maximise the beneficial change delivered by the project. The PMI[2] discuss stakeholder satisfaction and note that there are few concrete ways to measure it, instead for example, suggesting that in the case of a product's customers, their willingness to recommend the product could be is analysed, or a 'mood chart' could be constructed. These are however, fairly subjective measures. When the satisfaction of the project's key stakeholders is measured, one of the key groups that should be considered, is the project team itself. More tangible methods can be used to measure satisfaction here, such as for example, the unplanned turnover of staff. Whilst this is a far less subjective measure, it does not necessarily correlate closely with client satisfaction – the satisfaction of the project owner. It is customer or client satisfaction which is the subject here, rather than the wider view of stakeholder satisfaction.

Project success can also be a nebulous concept. It too will depend upon from who's viewpoint it is judged. In this chapter, it is viewed from the perspective of the project owner, the person or organisation that provides the funds for the project. The PMI[2] identify properly defined success criteria, or clear descriptions of what constitutes a successful project outcome, as being key to successful project delivery, noting that without them there is a risk of a lack of focus on project outcomes and hence poor project performance. P2M[3] reflects this too and document that definitions of project success should be included along with discussion of project benefits as part of the project's business case (see Chapter 3).

One way of measuring project success is by reference to the project's business case. The project was approved because the business case showed a positive gap between costs and benefits of sufficient size to make the project worth undertaking. Has the original excess of benefits over costs put forward in the cost benefit analysis been delivered? Has it been increased wherever possible and maximised by the project team? If this definition is chosen, then in order to protect the overall benefits defined in the business case, additional costs must be controlled, and benefits must be understood, managed and delivered.

12.2 How Is Are Benefits and Change Control Managed and Client Satisfaction and Project Success Defined Within a Project?

12.2.1 Benefit Management

Benefits are something that are often only focused on at the end of the project, when the delivery is complete, and it passes into the operational stage. They may not even be reviewed at this stage. The Iron Triangle devised by Dr Martin Barnes in 1969 defines the time, cost and quality (TCQ) constraints that project delivery is faced with. The meanings of T and C are straight forward within a project environment. The meaning of Q is perhaps less obvious, but it can be viewed as a measure of the suitability of the project's end result, the value of the project deliverables or the degree to which the project output is fit for purpose. These definitions mean that quality is measured according to the parameters of the project lifecycle, as opposed to the extended project lifecycle (see Chapter 4). In the discussion of TC and Q it seems as if benefits are scarcely mentioned. The benefits are discussed, as part of the construction of the business case (see Chapter 2), but the danger is that this is where they stay. Figure 12.1 below shows how strategy develops into net gains for the organisation through the extended project lifecycle:

It is notable that whilst TCQ describes what occurs in the project lifecycle phase, if the extended lifecycle is the focus, then rather than TCQ what becomes important is TC & Benefits. This is not surprising since a project benefit could be described as only occurring once an output has been produced and begins to be utilised, and hence a benefit can only be delivered in the in use phase, or extended part of the lifecycle. Noting this difference between the project lifecycle and the extended part of the lifecycle, it can be readily seen why projects might become fixated with delivering outputs rather than enabling benefits, thereby losing focus on why they

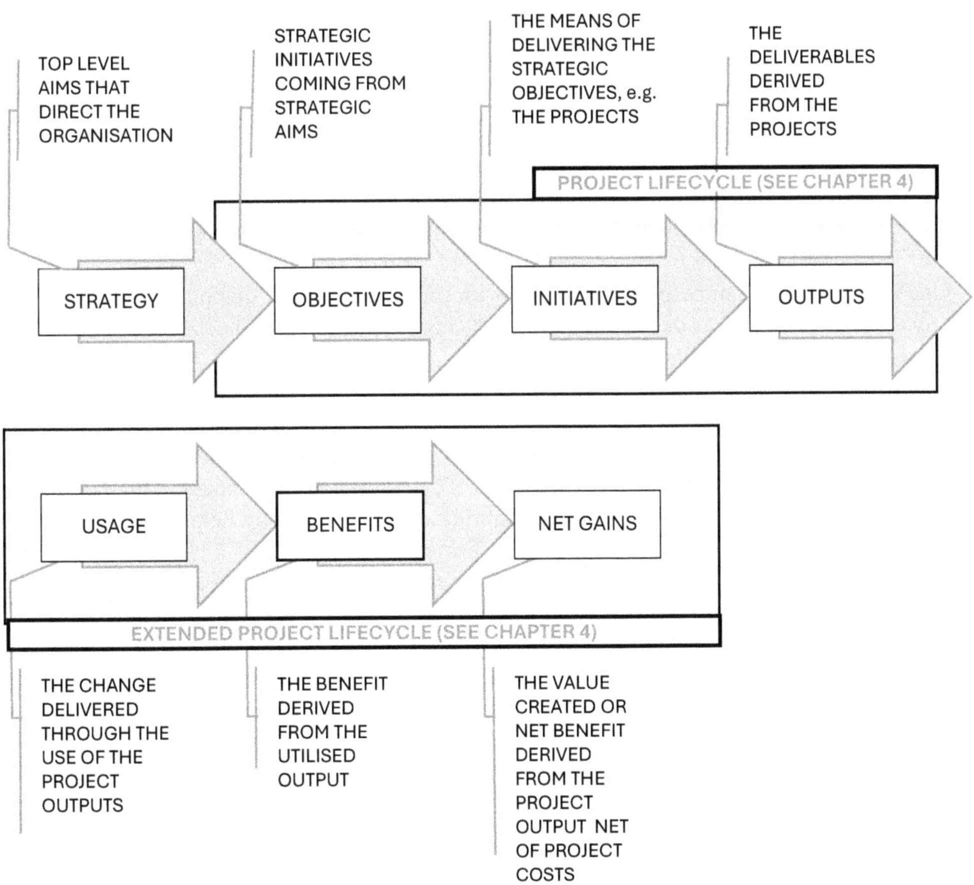

Figure 12.1 Organisational strategy to benefits and net gains through the project lifecycles.

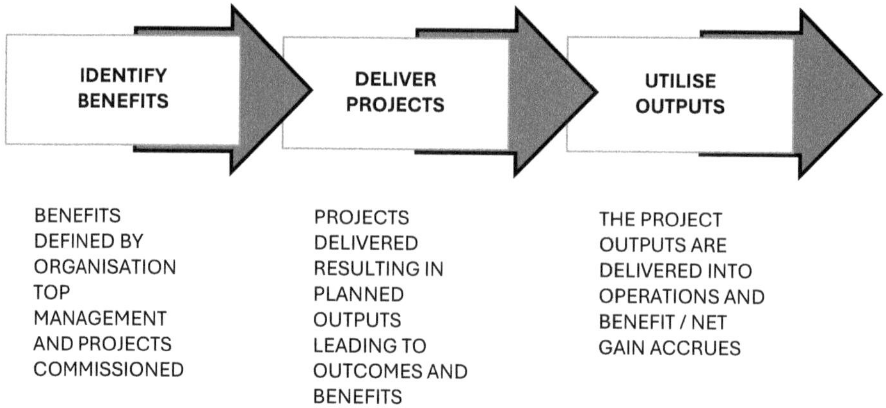

Figure 12.2 The three stages of benefit management.

were established in the first place, i.e. the business case and delivery of the benefits therein. It becomes vital therefore that benefits are managed through both the project and extended project lifecycles and indeed are the focus of project delivery.

The process for managing benefits has three key stages, described in Figure 12.2, within each of which there can be numerous challenges.

The management of the organisation need to consider which benefits are required in order to deliver the organisational strategy. Which projects will deliver which benefits must be considered, as should the question of how the projects will work together to deliver the organisation's overall benefits set. It is the more involved benefits that will require projects to deliver them, the simple, small benefits with a very local impact may not require a project in order to implement them.

One technique that can help align benefits with projects is benefit mapping, thereby moving the focus away from merely delivering to the TCQ constraints of the project and towards a focus centred on the project benefits.

Any organisation will have what amounts to a wish list of desired benefits. This wish list will help define the new 'end state' which the organisation aspires to reach. Each benefit will inevitably cost money in order to deliver it. If the cost of implementing the benefits is more than can be afforded, then some selection and deselection of benefits needs to be made. This can be done by plotting the paths to each benefit and weighting each path thereby helping to select which benefits should be delivered and through which projects.

An example of how this approach can work is shown in Figure 12.3 below which shows a fictious end state for a sales company that wants to deliver an increase in sales revenue.

Figure 12.3 shows the desired end state, Increased Sales Revenue, on the right-hand side. In order to achieve this though, numerous other benefits need to be delivered. These benefits are delivered through the three initiatives, or potential projects, shown on the left-hand side of Figure 12.3. The map shows how the benefits accrued through the deployment of the three initiatives deliver interim benefits before finally delivering the desired end state of Increased Sales Revenue. The question raised above though was which initiative(s) should be undertaken if the organisation cannot afford all three shown in Figure 12.3?

To help determine this, the contribution of each initiative to each of the benefits and hence the desired end state can be weighted as shown in Figure 12.4.

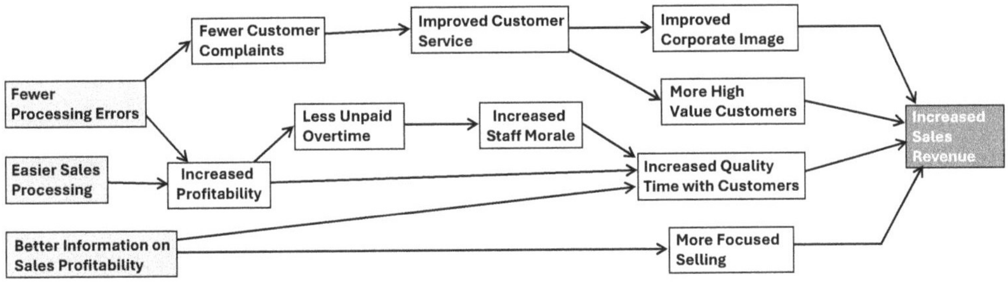

Figure 12.3 An example of a benefits map.

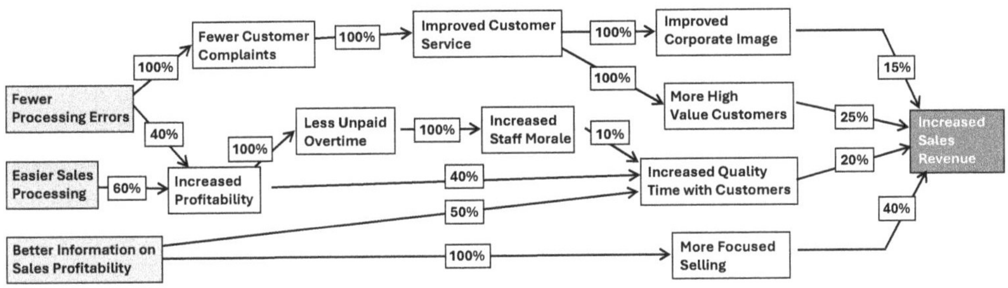

Figure 12.4 A weighted benefits map.

The contribution of each initiative can now be calculated. If the line between 'Better Information on Sales Profitability' and 'Increased Sales Revenue', at the bottom of Figure 12.4, is reviewed, then 'Better Information on Sales Profitability' can be seen to contribute 100% of the inputs to the benefit of 'More Focused Selling', which in turn contributes 40% towards 'Increased Sales Revenue'. 100% × 40% = 40%, so 'Better Information on sales Profitability', via this route, contributes 40% to the ultimate aim of increasing sales revenue. There is a second route between 'Better Information' and 'Increased Sales'. This can be seen to go through 'Increased Quality Time with Customers, providing 50% input to the 20% contribution to 'Increased Sales'. 50% × 20% = 10%. The total contribution therefore of the 'Better Information' initiative is 40% + 10% = 50%.

If this calculation is performed for each one of the three initiatives shown in Figure 12.5, then the relative value of each project or initiative on the left-hand side of the diagram can be estimated.

For the example provided, a clear ranking between the initiatives can be seen. Delivering 'Fewer Processing Errors' will contribute 43% of the overall benefits, delivering 'Easier Sales Processing' will provide 7% of the overall benefits, whilst the remaining 50% of the benefits are provided by delivering Better Information on Sales Profitability.

If finance for the project was in short supply, then Figure 12.5 would suggest that all things being equal, 'Easier Sales Processing' could be sacrificed and a majority of the benefits of Increased Sales Revenue could still be achieved.

The delivery of these benefits completes the second step, 'deliver projects' of Figure 12.2, the project(s) designed will have provided the planned project outputs that were designed to result in the planned outcomes and benefits. What remains of Figure 12.2 is part 3, the

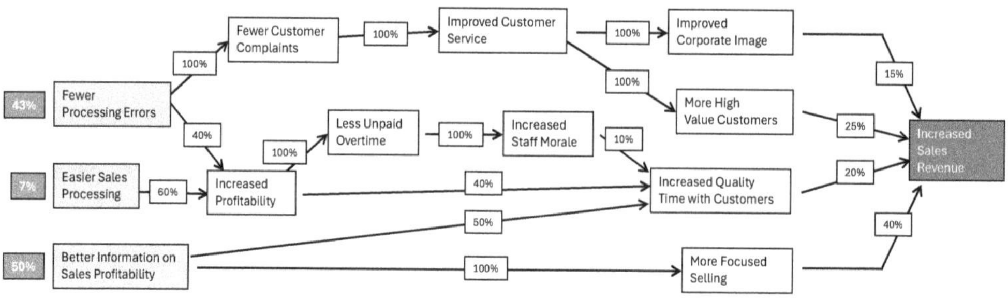

Figure 12.5 A fully weighted benefits map.

utilisation of the outputs. The requirement for the project now is to ensure that the outputs do result in benefits and net gains for the organisation. How this is done will depend on how long the project remains in service once the output is delivered into operations. An extended project lifecycle will naturally take account of the in service phase of the project. Even in the case of the shorter non-extended, project lifecycle, it is sensible, and normal, to not just end the project once all products are delivered into service, but to have a 'bedding in period' where it can be established that the products are fit for purpose and that the project sponsor is capable realising the desired benefits. Some of this will have been established in the end testing immediately before the outputs were set live, but there could be cause to check, for instance, that users of the outputs, wider than the test team, can derive the required benefits. This may mean revisiting the staff training, the ratio of outputs to staff, or even the finer points of how the project outputs work. This becomes part of Benefit Realisation, defined by the APM[1] as 'the practice of ensuring benefits are derived from outputs and outcomes.' A Benefit Review Meeting will be required once the project is live in order to assess the actual benefits delivered. The project team will have agreed the required benefits as part of the first step in the process described in Figure 12.2, and these will have been included in the map at Figure 12.3. These benefits (subject to any relevant change control) should have been delivered, and it is the purpose of the Benefit Review Meeting to establish whether this is the case. If it is found not to be the case, then action will need to be agreed to address the issue, and an additional follow up Benefit Review Meeting planned.

The benefit mapping process is therefore useful in both describing and understanding the interactions and relative importance of proposed benefits, and also in providing a basis for reporting the projects ongoing progress.

12.2.2 Change Control

The answer to the question of whether changes in projects are a good thing or not, is either 'yes and no', or alternatively 'it depends'.

It depends firstly on whether they increase the value derived from the project – the gap between costs and benefits first described in Chapter 2, and then secondly on whether they are introduced in a controlled manner or not.

If a project is subject to a very large number of demands for changes, then it potentially may be that the wrong, or poorly defined project is being delivered. This will need review. Even a well-defined project however, can require changes to be made through its delivery lifecycle. Suppose a project has a 10-year delivery period, then the environment it delivers into could be significantly different to that envisaged at the start of the project. If benefits are to be gained it may well be that changes in the project's output characteristics must be made in order to protect and ultimately deliver the benefits. Uncontrolled changes though can be disastrous for a project. They are at the root of what is termed 'scope creep' where the list of requirements for a project appear to grow exponentially until the project becomes impossible to deliver and the project team itself is tagged with failure. To avoid this a project needs to define from the start how it will manage requests for change. A formally agreed and signed off business case is invaluable here as is the benefit map described above. These documents provide a baseline against which any potential changes can be assessed and authorised. There are a number of possible causes of scope creep, many of which are listed in Table 12.1 below, but if a project change control process is established and adhered to, the excesses of scope creep can be avoided.

In order to accept only those changes which can enhance project outcomes and hence benefits, a formal change control process must be established.

A key document that feeds into the process shown in Figure 12.6 is the change control form. Its purpose is to document the requested change and to evaluate the impact of the request on project cost and benefits. An example change control form is shown below in Figure 12.7.

The form has four key parts, shown as A, B, C and D in Figure 12.7.

Section A provides details of where the change request has come from, and a full description of the change required. It is important that as much detail can be provided as possible in order that it can pass onto full analysis in Section B of the form. Section B reviews the impact on the TCQ components of the project should the change be accepted. The current baselined TCQ figures will be included in the original business case plus any subsequent agreed updates to it. In assessing the impact of the change what needs to be gauged is the gap between the cost and benefits delivered by the change compared to the gap that was accepted and recorded in the

Table 12.1 Possible Causes of Scope Creep

Possible Cause of Scope Creep	*Potential Reason*
Lack of a change control system	Poor initial project set up and understanding of how to control project dynamics
Poor channels of communications	Poor initial project set up and understanding of how to control project dynamics
Requirement gaps	Project requirements are not fully or accurately gathered
Requirements are present, but poorly defined	Requirements should be described in measurable ways and should not be subjective in nature
Some important stakeholders have not been included	Stakeholders were incorrectly analysed and missed in the early stages of the project
Customer expectations	Customers may assume that outputs are included when they are not. This suggests poor project baselining
Lack of relevant experience in this type of project	The assembled project team may not reflect the skills or attributes that the project requires
Poor project management	This could be due to the project manager or other senior members of the project team failing to employ, or trying to subvert, due project processes

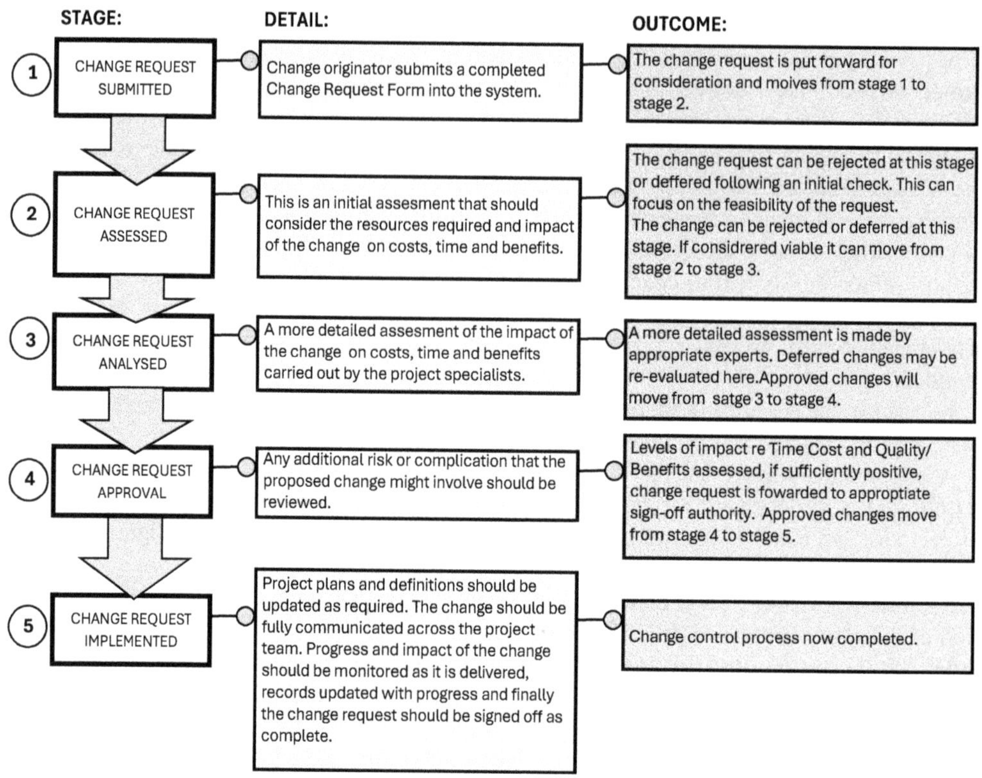

Figure 12.6 A change control process.

latest, baselined, business case. The change is worth contemplating if it will increase this gap. If the gap is increased, then the impact on the client should be assessed in terms of the viability of spending the additional monies required to realise the change, and then the desirability of doing so given the size and nature of the increased benefits that would be delivered. The outcome of these deliberations should be recorded in Section C where the change is formally accepted or rejected by the appropriate change authority. Finally in Section D, if the change is approved, a record signed off by the project manager should be made of the implementation of the change and of the changes full incorporation into the project delivery.

What is key is that a structured decision is made taking full account of as much relevant information about the costs and impacts of the proposed change as possible.

12.2.3 Client Satisfaction

Client involvement in the change control process will help to ensure that the gap between costs and benefits is maximised. It is particularly important to achieve this in the eyes of the client as they are the ultimate arbitrator of the value of any additional benefits which the project may deliver.

Morris[4] equates the success of a project, and hence the implied client satisfaction, with the achievement of the projects success criteria, enabled by critical success factors, as addressed in the project's business case (see Chapter 2). The APM[1] support this saying that project success relates to the satisfaction of the stakeholder needs as defined by the success criteria agreed at the

CHANGE CONTROL FORM			
SECTION A:			
PROJECT NAME:		CHANGE No.	
DETAILS OF CHANGE REQUESTED:			
CHANGE REQUESTED BY:		DATE REQUESTED:	
SECTION B:			
CHANGE REVIEWED BY:		DATE REVIEWED:	
IMPACT ON OTHER PROJECT COMPONENTS:			
IMPACT ON PROJECT **BENEFITS** (INCL. **QUALITY** OF PROJECT OUTPUTS):			
IMPACT ON PROJECT **COSTS**:			
IMPACT ON PROJECT **DELIVERY TIMES**:			
SECTION C:			
REJECT CHANGE		APPROVED BY:	
ACCEPT CHANGE		DATE APPROVED:	
SECTION D (complete only if change request is approved):			
CHANGE IMPLEMENTED BY:		DATE IMPLEMENTED:	
AGREED BY PROJECT MANAGER:		DATE AGREED:	

Figure 12.7 A Change Control Form.

beginning of the project. This agreement should be signed of between the project delivery team and the sponsor, or client, of the project.

The fact that the success criteria are agreed by the client does not of course mean that other stakeholders needs are excluded from the definition of them, as delivery of some of these other needs, can also directly impact the achievement of the client's project satisfaction.

Testing of project outputs will occur throughout the project lifecycle, but one key part of the testing regime is the end testing, or the testing of the completed product just before the project is set live. This is the stage where end users of the project outputs need to be involved in order to establish that the products match the requirements of the business case and to confirm that the project outputs are capable of delivering the required benefits once the project deliverables start to be used. If there is a shortfall between the benefits required and the capabilities of the project, then corrective action followed by another round of testing may be required.

The involvement of the client and other users of the end products at this stage will help to ensure client satisfaction and assist in delivering the positive difference, or gap, between the project's costs and its benefits outlined by the business case. Once the testing is completed, the outcome should be documented by the project manager through the completion of a formal test report. This should give full details of the outcome of the testing and should be signed off by all those involved, particularly the end user who undertook the tests.

12.2.4 Project Success and Benefit Reviews

Project Success is a difficult concept to define. It can depend on which stakeholder is asked. It can be measured from the viewpoint of how the project itself was run, or it can consider if the project itself was successful in achieving what it set out to do, in other words, was it the right project.

Probably the most frequently used measure of project success is through direct analysis of project outputs in terms of the TCQ parameters presented in the business case. This establishes the project sponsor or client's perspective as key in determining project success. It is an understandable perspective since this is where the project's funds come from, either directly or indirectly. It is however often the case that a project has two or more key stakeholders. Consider the case of a project focused on the development of a new commercial undertaking. The project sponsor may raise funds for the project through bank finance, which will mean that the bank will also have some success criteria for the project, as may staff employed on the project concerned with the development of their careers, or the end users who will actually directly use the project outputs. The example can be expanded. If the commercial endeavour were to be a multi store shopping complex then the end users could be defined as each of the companies that rent a store in the complex, or could even be each of their customers, or each of their employees. The number of stakeholders could be seen as being never ending as, for example, more customers and more employees are added over time, in which case the success criteria will develop and change over time. Project success under this definition of stakeholder becomes impossible to judge due to the myriad of, often conflicting and changing views that develop.

As mentioned in Section 12.2.1 above, the primary purpose of the project is to deliver project outputs to the sponsor and to enable the sponsor to realise these benefits so as to maximise the gap between project costs and project benefits. Project benefits delivered to the project sponsor at the end of the project lifecycle will include an assessment by the sponsor of the valuation of the future benefits that will be reaped. Benefits therefore become less tangible since they are yet to materialise, and what is being evaluated is the ability of the project outputs to deliver the defined benefits, once the project's delivered qualities begin to be utilised. This equates to project success being measured through consideration of the TCQ parameters discussed above.

Project success is therefore reflected in the maximisation of the positive gap between costs and benefits detailed in the business case, the achievement of project delivery according to the TCQ constraints, and the subsequent ability of the organisation to achieve the desired benefits.

In order to monitor project success and the achievement of the benefits some degree of benefit realisation reviews should be established. The project, if operating under a project lifecycle, (as opposed to an extended lifecycle) should establish a benefit review system to ensure benefits are delivered or are capable of being delivered in full, post transfer to the client.

Once the outputs of the project are delivered, a period of time, sometimes known as a 'bedding in process', should be allowed for these outputs to ensure that benefits can be harvested and fully realised. An IT project for example, may deliver perfectly usable outputs but may neglect to train the end users on how the systems should be used thus negating the opportunity to deliver any benefits. A bedding in period or series of benefit review meetings will be able to detect, and correct, this omission.

It was also said at the beginning of this section that project success can also be viewed from the viewpoint of how the project itself was run. This is often included in the project review at the end of the project lifecycle. Here lessons can be learnt for future projects by assessing whether or not the project management functions, such as those discussed throughout the book, worked in such a way as to help enable or hinder the project's successful completion.

If both the processes worked well and the benefits are seen to have been delivered to TCQ, then the project can be regarded as a success.

12.3 Project Review: The Impact of Changes on the Berlin Brandenburg International Airport (BER) Project

12.3.1 The Project Background

It has been said[5] that the idea to build an airport, to be named "Berlin Brandenburg International" (BER), in the German capital following the country's 1991 reunification was flawed from the start. In the 1990s it was clear that following unification the German capital would grow, but it was also recognised that the site in the Schönefeld area near Berlin, would not be able to expand in the same way. Other locations, situated further away from the capital, but still close to Brandenburg, that could more readily accommodate the runways and expected passenger volumes were put forward, but the original site selection was not modified.

12.3.2 The Project Issue

The project eventually got underway late in 2005. By this time the government and the federal states of Brandenburg and Berlin had determined to construct the airport themselves at the original site following the failure of negotiations to let construction contracts to the private sector. In the early days of the project, it was claimed that the building would be too small to handle the predicted volume of traffic. The airport's project management team requested that the airport be redesigned and that the 30 October 2011 opening date be postponed. One of the project engineers and airport planners was quoted[5] as saying that the decision to continue with the original plan, despite it being too small in order to meet the original completion date was very much a mistake. Inevitably, around 2007/8, it was formally recognised that the airport was indeed too small, but valuable construction time had by this point been lost. In 2010, new EU safety regulations meant that additional work, relating to the expansion of the airport's security screening area was required.[6] EU directive 297/2010 required all EU airports to phase-in the deployment of security related detection methods by 29 April 2013. These were designed to provide the capability to screen liquids, aerosols and gels to ensure liquid explosives were not inadvertently loaded onto planes. This caused the German Federal Police to require a doubling in the space given over to screening activities. The reason for this change was made public at the time, whereas it was suspected[6] that other causes of delay were not made so publicly available.

The impact of the EU directive helped, at the time, to delay the predicted opening ceremony by some nine months to June 2012. The press however reported that in addition to the altered EU safety requirements, it was also the building and security technology, and the fire protection systems that required changes to maintain functionality that caused the delay. A large number of change requests are noted in the working paper on the BER airport published by the Hertie School of Governance.[6] It is reported that not only were tenders undertaken before detailed design was complete, but tenders were also undertaken only to be significantly changed subsequently. This inevitably resulted in nugatory work being undertaken and claims being made by contractors as the required designs were subject to significant change.

It is worth noting that the impact of the EU directive is cited by the Hertie Report[6] as being indicative of a 'rushed, thoughtless and un-monitored detailed design process' which is quite an inditement of the project planners and would, inevitably, have led to multiple change requests as the design matured.

The changes resulted in design activities running in parallel with construction activities. This is not uncommon within construction projects, but given the size of the changes being requested, the enormity of their impact was felt throughout the project. The ongoing design changes had knock on impacts in areas such as the fire safety and smoke extraction systems which needed to be redesigned to reflect, and to operate safely, within the expanding building dimensions. Bosch GmbH are reported as being one of many contractors responsible for components in the smoke extraction system and the building changes meant that they were faced with over three hundred change requests. The number of changes requested clearly impacted not just Bosch but other contractors too. The airport planners were faced with multiple change requests from the airport authorities, FBB (Flughafen Berlin Brandenburg GmbH), and these changes were reviewed in terms of their impact on project costs and task durations. Four hundred and eighty-seven changes were accepted in relation to the design of the passenger terminal alone in the period up to the early part of 2012. Prior to this, the chaos caused by the sheer number of requests and the interconnectivity of various parts of the airport build had meant that at the end of January 2010, a stop was put on any additional change requests by FBB.

The ban on additional changes however was not altogether successful. A graph[6] shows that in the period between January 2010 and June 2012, immediately post the stop on any additional changes being mandated, the number of changes requested far from stopping, actually increased when compared to the period between October 2007 and January 2010 directly before the 'stop' order was issued.

Ultimately, the airport opened on 31 October 2020[7] some nine years after the initial planned opening and some three decades after the plans were first discussed and more than €4bn (£3.6bn) over budget.[8]

12.3.3 How Could a Developed Change Control Process Have Helped?

Changes were multiple and the impacts were complicated.

A change control process was evident in places, the planning authority for the airport, pg bbi,[6] did assess the impact of proposed changes relating to the passenger terminal, but it is unclear if the overall impact on the project as a whole, rather than just the work package directly involved, was sufficiently reviewed as part of this process.

It is also doubtful that the full impact of not delaying the opening and redesigning, as called for by the airport's project managers,[5] was understood. It is also unclear if the option of delaying the opening and redesigning was at the time run as a change request and compared to the as is scenario.

The change control process seems to have been run potentially at a 'work package' level meaning that the proposed changes were judged only at a relatively local level in terms of the WBS (see Chapter 8), and the impact on the project overall may not have been calculated/envisaged. The number of contractors involved in just one area of the project, such as the smoke extraction system would have needed a centralised communication system to have kept all contractors up to date with approved changes and a developed change control system to ensure that the effect of any changes was understood across all contractors and across the airport as a whole.

The need for a change control process is clearly evident, but it must be one which fully understands and reflects the project. It is clear that the BER airport was subject to a number of issues such as political pressures and inflexible timescales alongside the problems caused by the changing deliverables. Having a change control process in place is insufficient in itself. It needs to be one where the hierarchy of change request approvals is fully documented, approved and understood. It has to be one which facilitates a full understanding of the impact of changes across all parties and is realistic in its decisions, based upon a solid project baseline. These factors should be listed as part of the project process definition documents.

12.3.4 Conclusions on the Berlin Brandenburg Airport Project

The issues surrounding the airport construction were summed up by the then chief operating officer of FBB, who said that the project's problems were the result of a 'whole range of changes in planning and use' resulting in 'constant changes' in design. The lack of quality planning and late delivery of design documents added to the problems, and was directly responsible for increased time pressure and delays within the project. The lack of certainty generated by these factors bore heavily on the project's level of success. Clarity of purpose is key to the definition of any project's goals and the successful achievement of them. Changes will be necessary in almost any project, but it is the way in which they are managed and specifically the ability to implement a change control process in a realistic and meticulous manner that will help enable the project to be successfully delivered.

12.4 The Management of Benefits, Changes, Customer Satisfaction and Project Success Before the Birth of Modern Project Management*

(* In this section, the research relating to the construction of HMS Barham 1912–1915 mentioned in the introduction is drawn upon to help illustrate how the subject of this chapter has developed.)

Section 12.4 continues the review of project practices prior to project management's formal codification. In the study of the construction of HMS Barham 1912–1915, it is notable that the discussion of benefits is very limited. The project was delivered under a 'project lifecycle' not an 'extended project lifecycle' (see Chapter 4) and so was primarily concerned with the delivery of capability rather than benefits. The required capabilities were set out in the design specifications and drawings first authored at the start of the 1912 Shipbuilding Programme.

Change Management was in evidence, the assessment or recording of which is shown in Photo 12.1 below, which lists 'changes' as 'extras' on the right hand page.

Photo 12.1 shows details of headings such as a top-level description of the additional requirement, a reference to the 'Letter Book' in which the change would have been discussed in more detail, columns headed 'amount submitted' and 'date submitted', are followed by details of the date approved and then by columns detailing the date and amount that was finally paid in relation to each 'extra' that was approved. The journal carefully records the detail of each change required and provides a record of the overall sum submitted, following an estimation process for each change. Some of the changes listed on the left hand page, include providing a search light platform on the aft funnel, providing extras in connection with oil driven generators, and seemingly quite small items such as the substitution of kit lockers for racks.

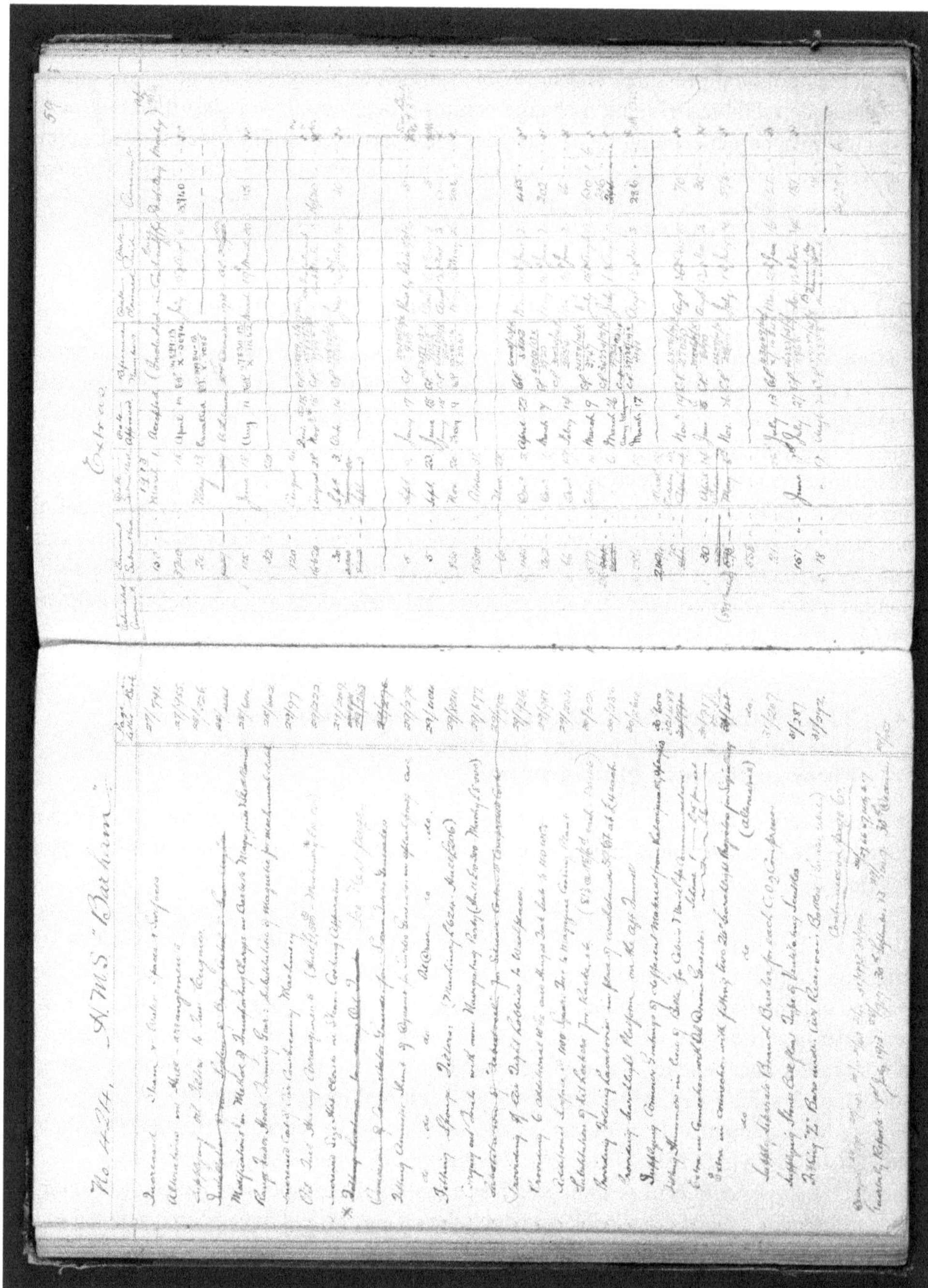

Photo 12.1 An extract from the 1913 'extras' entries relating to the construction of HMS Barham. By permission of the National Records of Scotland and University of Glasgow Archives & Special Collections, Records of Upper Clyde Shipbuilders Ltd, Clydebank Division collection, GB248 UCS 1/64/1.

Changes can be seen as being accepted once the cost has been submitted, but also being rejected at this point. The column headed 'Date Approved' shows that the third entry relating to the supply of oil filters at a cost of £20 was cancelled, whilst the fourth entry is shown as being 'withdrawn'. The majority of entries however are shown as being approved and the date of this approval is recorded. The emphasis in the journal is very much on the costs approved and traceability of the requested design changes, the impact on project duration and quality of output is less obvious. This therefore shows the key commercial aspects of the changes as might be expected from a contractor's records.

The focus on the remaining two TCQ parameters, time and quality, can perhaps be seen more clearly in the documents relating to customer satisfaction shown below at Photo 12.2.

An extract from the letter in Photo 12.2 reads 'it is desired to invite attention to the satisfactory performance of Messers. John Brown & Co. Ltd. In the matter of time taken by them to build destroyers.' Whilst this is perhaps a little understated, it does demonstrate that the company were able to deliver ship construction projects in a timely manner and to the Admiralty's satisfaction. The second document shown in Photo 12.2 indicates that of the fourteen constructors listed, John Brown built twenty-seven ships for the Admiralty between 1914 and 1918, which was the largest number of Admiralty vessels by any one constructor and 22% more than the output of Thorneycroft, its nearest rival in terms of vessels completed.

The completion certificate for HMS Barham was signed by the Admiralty on 28–9–15 prior to her commissioning on 19-10-15, and post sea trials. This was a detailed sign-off against ship dimensions and materials thereby accepting the ship as completed. The acceptance of the ship into service by the Admiralty would have been validation of the quality of the vessel, and the restrained praise for John Brown's timely delivery of ships is evidence that the ships were delivered on time. The fastest ship built for the Admiralty by the fourteen constructors was the

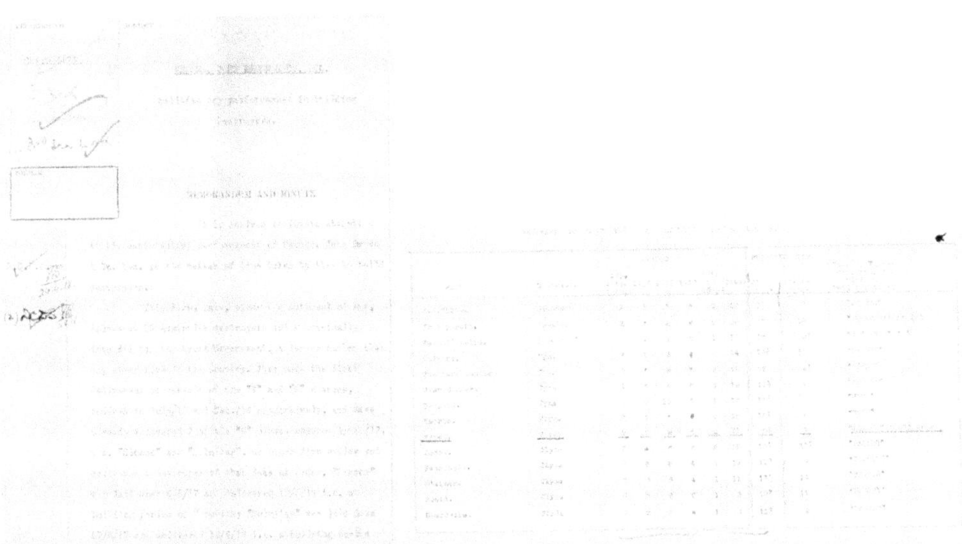

Photo 12.2 An extract from a letter from the Admiralty to John Brown & Co. congratulating them on their shipbuilding efforts. By permission of the National Records of Scotland and University of Glasgow Archives & Special Collections, Records of Upper Clyde Shipbuilders Ltd, Clydebank Division collection, GB248 UCS 1/9/193.

John Brown built HMS Simoom, (as shown in Photo 12.2) an S-class destroyer, launched on 26 January 1918 which served as part of the Grand Fleet. It was launched in just seven months, one month quicker than the next faster to be built, the HMS Moon also a destroyer which is shown as being launched in 1915 on the Clyde by Yarrow.

Overall, it can be seen that changes requested, and required, during the building of HMS Barham were delivered in a controlled and structured manner. The processing of the change requests meant that the desired capability was signed off through the tests carried out and that this resulted in the ship being accepted into service as witnessed in subsequent letters of approval sent directly from the Admiralty to the constructor.

References

1 APM. (2019). *APM body of knowledge* (7th ed.). Association for Project Management.
2 Project Management Institute. (2021). The standard for project management and a guide to the project management body of knowledge (7th ed.). Project Management Institute, Inc.
3 Ohara, S. (2017). *P2M – A guidebook of project & program management for enterprise innovation* (3rd ed.). Project Management Association of Japan (PMAJ).
4 Morris, P.W. (1994). *The management of projects*. T. Telford.
5 Schlote, H., Lowe, J., Williamson, E., Syn, J., Wood, M., & Planck-Prideaux, L. (n.d.). *Grounded: A management assessment of the complex Berlin Brandenburg airport project*. https://harryschlote.com/projects/airport.pdf
6 *Between ambition and realities*. https://www.hertie-school.org/fileadmin/2_Research/2_Research_directory/Research_projects/Large_infrastructure_projects_in_Germany_Between_ambition_and_realities/3_WP_BER_Flughafen_Berlin_Brandenburg.pdf [Accessed 6 December 2024].
7 Haack, J. (2021). *100 days later: The first three months of operations at BER Airport*. https://www.internationalairportreview.com/article/154517/first-months-operations-ber-airport/ [Accessed 15 December 2024].
8 Connolly, K. (2020, October 7). *We were a laughing stock': Berlin airport finally finished as Covid bites. The Guardian*. https://www.theguardian.com/world/2020/oct/07/berlin-new-airport-covid-germany-budget-late [Accessed 8 December 2024].

Chapter 13

Business Readiness and Project Closure

Once a project reaches its conclusion, two things must happen. The outputs from the project need to be transferred into the business-as-usual domain and the project activities must be closed down.

Part of successful project delivery is the maximisation of value for money represented by the project. Efficient delivery means that costs must be watched at all times, and one place where there is potential to take the eye off the project costs is towards the end of a project where everything looks to be going well, and success appears guaranteed. This is caused in part by the effect examined in Chapter 7, where overestimates can become self-fulfilling prophecies as the amount of head room on a budget causes the project team to relax and start being less diligent in their tracking of the project financials.

By the time a project closes there is another key component of project success that must be delivered. If the project is to be deemed a success, it is key that the business itself is ready to make the best use possible of the project outputs thereby maximising the project's rewards, or benefits, and this element of project success is captured in the phrase 'business readiness'.

13.1 The Meaning of Business Readiness and Project Closure

The APM[1] describe the purpose of business readiness as ensuring the transition into use, by the temporary project team, of the project outputs, which will in turn enable the permanent staff of the organisation to fully benefit from the changes delivered. Project closure is recognised as a necessary skill capable of impacting projects through either a planned or unplanned (early) end to a project. The PMI[2] identify 'closing' as one of the five project process groupings used to organise and deliver a project (the others being initiating, planning, executing and monitoring and controlling). The view of the importance of project closure expressed in P2M[3] is very much in line with both the APM and PMI and it stresses that 'the closing process formally recognises project completion and transfers the project to an organisation that will effectively control and manage it using the deliverables and created values.' P2M recognises it as a formal process that requires a formal sign-off which in itself defines the process, and indeed the project, as complete.

13.2 Managing Business Readiness and Project Closure

In order to ensure that the business is ready to take full advantage of a project's outputs, some specific project tasks will need to be completed.

The work to make sure that a business is ready to utilise a project's output begins early in the project lifecycle. Firstly, the project team should understand how the outputs will affect

DOI: 10.1201/9781003470748-14

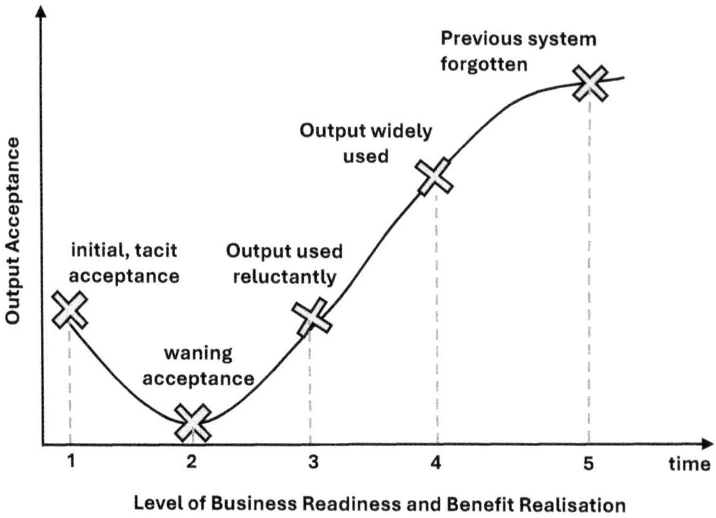

Figure 13.1 Model of project output acceptance and business readiness/benefit realisation.

the business-as-usual activities of the organisation. Functions and processes rarely happen in isolation so aspects of interoperability between business functions need to be considered. If for example, a new IT system is to be launched, how will the interfaces with the financial reporting system work? How will the new system operate alongside the existing sales system? The mechanical interfaces should have been specified in the product specification, but what about the softer, people related interfaces? Do staff know and understand the different ways in which the new system will work? Have they been made aware of any process changes and altered methods of working that may be required? How a team typically adapts to the challenges facing them as the project is set live is illustrated above in Figure 13.1, which has similarities with the Katzenbach and Smith model of team effectiveness shown in Figure 6.11, but focuses on the degree to which the business is ready to accept new working methods.

Figure 13.1 illustrates that initially the concept or idea of a new system, component, or way of doing things, may be tacitly accepted (point 1 in Figure 13.1). There may be an element of denial in this, a view that 'this will never happen" or 'this won't affect me so I don't need to worry about it', driven by the feeling that the change being implemented is intangible and won't be in action for some time, and hence, the impact does not feel real. As time passes, point 2 is reached where users realise that they are going to be faced with a change to the existing system, which while it may have had its faults, they were known, understood and work arounds were established. There is a reluctance to accept the reality of the changing system, this is however, a natural part of the acceptance of change, almost a grieving process as seen in other change management soft issues, and it is an important part of the acceptance process. Point 3 represents the position where the project output begins to be used but is not fully accepted by those using it. It may be that it is used in isolated areas with other parts of the organisation finding it difficult to let go of the previous system. Gradually levels of acceptance increase as point 4 is approached until eventually point 5 is reached where the new system has become 'the system' and previous approaches have become forgotten and project benefits become fully realised.

The job for the project team is to move the organisation as quickly as possible from point 1 to point 5. This has a direct impact on the success of the project as if benefits can be gained from the deployment of project outputs at an early date, then the outputs will be producing benefits sooner and for a longer period. More benefits will therefore be gained before the project deliverables are retired. If more benefits can be obtained, then the gap between project costs and benefits as described in the business case (Chapter 2), will be greater. There are various steps that can be taken to speed the journey from point 1 to point 5. The steps are centred around ensuring ownership and familiarity of the end users with the outputs and so these measures should begin in the early stages of the project lifecycle and not just appear at the end of it. The involvement of end users in the definition of the problem that the project is addressing is vital, as is the testing of the developing project outputs designed to address these problems. It is important to not merely seek agreement with the any predefined project views, but rather, it is key to hear and to consider any opposing viewpoints rather than immediately seeing them as a resistance to the changes being implemented. This early engagement will allow greater alignment between the project's end users and the project's end results, and will in addition highlight any potential gaps or misunderstandings in the project's definition. One project task that should however be considered for inclusion toward the end of the project, once the system has crystallised, is the end user training. When the project output is steady in terms of development, and ready to be set live, the project needs to ensure that all end users, not just those engaged in the earlier project phases, are equipped and able to use the new system. Without this the project benefits are at risk, not only because costly errors could potentially be made in using the project deliverables, but also because if the launch is problematic, users can begin to develop an inbuilt hostility or resistance towards the new methods delivered by the project.

There are some key considerations for the project manager to consider in advance of a project being set live.

A project very rarely affects just one area of a business. A cross-functional readiness meeting or team (dependant on size and complexity of the changes) should be commissioned to review the impact of the project across all areas of the organisation that could potentially be impacted. There will inevitably be points were output from the new system will handover to other parts of the organisation and here it is useful to hold a forum, building on earlier project work, where an understanding of the impacts of project outputs can be fully analysed, focusing on ensuring that change in one function does not adversely impact another.

Testing of the end product's ability to perform as specified is a key stage in any project lifecycle. What should also be considered alongside functionality testing, is stress testing and contingency planning. Stress testing considers how the new component or system works with high volumes of demand. For instance, a new online quotes system providing quotes to multiple third parties may experience extra heavy demand at certain times or given certain circumstances. Will the system be able to handle the higher level of demand and if not, then what contingency plans can be put in place?

When the new system, be it IT based or not, is set live, will this be done gradually, a phased release, or all at one go, a 'big bang' release? The answer to this question may depend on many factors such as the size of the release, the number of operations impacted, or the nature of the operations being impacted. Some projects can't be released using a phased approach, whereas some operations may consist of multiple parts which could be updated one by one. How this part of the project is tackled often comes down to the related assessment of risk. A phased release may well enable an issue to be ironed out without the danger of impacting the whole organisation. A project's releases could be phased across different functions of an organisation, for example if a head office was to be relocated then the infrastructure could all be ready, but the organisation could potentially move one function at a time, the HR department followed by the Finance department, followed by the Legal department etc until all departments have been relocated. This would be lower risk than moving all departments on the same day. Alternatively, the project release could be phased geographically. Suppose a new working method was to be introduced across a series of regional offices, then the release might first be delivered to one office in the Southern Region then, if successful, to all the office in the Southern Region, followed by the Eastern. Western and Northern regions. Both of these approaches would lower the risk and allow for implementation lessons to be learnt.

Contingency planning is another consideration. What will happen on day 2 of a launch if the system falls over and stops working? Can the system be backed out and the original restored if needed? What tasks would need to be prerformed to enable this? Which staff would need to be available? In essence the fall-back position needs to be determined in case the launch meets an unpredicted issue. Thorough testing of the project outputs will help mitigate this risk, but the possibility, however remote, will remain.

While training has quite correctly been mentioned as a vital pre go-live task, there will be some staff within an organisation who don't need training on the new output but will need to be informed of the timings and other details of the changes planned – another business readiness task for the project manager to schedule.

Once a date has been confirmed for the project to be set live, then detailed planning of the event is required. It maybe that a far more detailed go-live timetable is required, scheduling activities not just on a daily basis, but on an hourly or even per minute basis. Numerous issues might need to be considered dependant on the type of project. The project might be best set live after office hours, or even at midnight when there is less activity for the new service to handle, but still some hours left after post go-live tests are completed in case of problems needing to be addressed. Will communications be needed between different sites or different groups of project staff, if so, how will this be organised? Will the go-live require observers in position to report back, or a series of floorwalkers to help users use the new system? Will remote helplines be needed in order to provide user support? How will each department report issues – will there be a go-live organisational chart showing how this will be handled for each department, or will it be a single point? How will issues be escalated?

Of course, how complicated this process is will depend upon the environment into which the new product is set live. The process relating to the completion of, for example a single new build house is likely to be far simpler than that needed to support the launch of a new IT system across all branches of a multi-national company.

Business, or end user, readiness is key to the successful launch of any project. Issues such as those discussed above therefore become crucial considerations as project outputs begin to be transitioned into service and as the project approaches closure.

Business readiness though is not the same as project closure, it can be prepared for throughout the project lifecycle by, for example, including end users in the design of project outputs at an early stage, and by inclusion in the test design, delivery and sign-off at later stages.

Project closure occurs, and can only occur, as the name suggests at the end of the project and can be approached as a mini project in its own right. The top-level breakdown could look something like that shown in Figure 13.2, noting however that every project will be different, and that the business readiness tasks although part of the review of project tasks, are accounted for in detail elsewhere:

The review of project tasks is fundamental and will check that all project tasks agreed with the customer as being in scope have been successfully delivered. The training element discussed above should fall within these, as should the transition of project outputs to the in-service

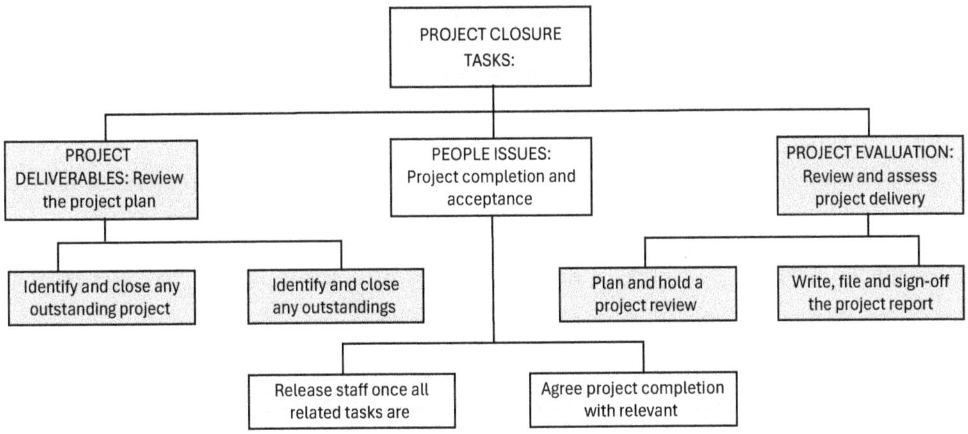

Figure 13.2 Top level project closure tasks

environment, but if any tasks are found to be outstanding, then a further detailed plan for completion of these will need to be enacted. Admin tasks will embrace the completion of any project documentation, including any handover documentation that the client may require. All contracts should be reviewed to ensure that they have been successfully concluded and terminated and that all project dues have been paid. Final client acceptance sign-off of the project outputs should also be obtained as a record of their satisfactory delivery.

There are a number of closure issues that relate directly to the people who have been involved in the project. Key amongst these, are those who have worked directly in the project team itself. The project team, once their job is complete, will need to be released from the project so that they may be assigned to other projects. If this is not done efficiently then there is a danger that costs will continue to accrue to the project. This process should also include contractors, suppliers and other partners who have been involved in the project. All should be informed that the project has been completed and the project manager should confirm and settle any final payments due.

The third of the key areas to consider when closing a project is the final project evaluation, or post project review. This will consider two key areas, the success of project delivery and the collation of lessons learnt which could improve future project delivery.

The success of the project should be reviewed and recorded in the project's final report, produced after a project evaluation meeting has been held. The report should appraise the time, cost and quality aspects of the project deliverables as laid out in the most recent iteration of the project's business plan. The review shouldn't be limited to an assessment of time, cost and quality, but should also consider other aspects such as team learnings and team successes. It is often best to keep issues such as team learnings, which often come from errors or mistakes, generic rather than specific, but team successes can have a more specific focus. A meeting should be held as close as possible to the project completion date to prevent facts becoming blurred, or forgotten, with the passage of time. A pre-meeting questionnaire can help in capturing some of the key points. Questions might be asked such as:

- Which three things do you consider went well within the project?
- Which three things did not go so well?
- What should be done differently in order to improve future projects?

The PM or moderator of the meeting can develop some of these points by discussing points such as:

- Where there any particular issues with timelines set by the project plan?
- What were the difficulties that blocked progress?
- Were there sufficient resources to complete the project by the deadline set? If not, what additional resources should have been provided?
- If the project were starting the project again, what should be done differently?
- Should the project be considered a success? What are the factors that lead to this conclusion?

A key part of a project review, often run separately, but none the less an essential feed into the project evaluation, is the benefit review appraisal as described previously in section 12.2.1. A review of the realisation of benefits as laid out in the project's business plan, along with a review of the projects critical success factors should feed into points raised at the project evaluation meeting.

Once the report has been issued one of the key challenges for a project organisation is to ensure that the lessons learnt are indeed learnt and used to facilitate future projects rather than becoming merely 'lessons identified'.

13.3 Project Review: The Impact of Business Readiness on the Project: the Poll Tax

This is a slightly unusual choice of project. It is firstly a massive project in the sense that it was deployed across the whole of the UK, and it was also a purely political project, without the production of a physical end product, whereas the majority of other case studies included have been corporate projects, or have had corporate links, of one type or another. The impact of the project was also so large that it has been credited with helping to bring down the UK government.[4]

13.3.1 The Project Background

Prior to the roll-out of the poll tax in 1990, local government services had been funded through a system of local rates.[5] The existing tax was levied on both residential and commercial properties based upon a notional historic estimate of the property's rental income. This charge was fixed irrespective of the number of people living in the house and hence their implied usage of the local services, meaning that a property having a lower rateable value paid less than a larger property with fewer residents. Margaret Thatcher, the UK's then Prime Minister, and her government considered this to be inequitable and thought that it did not reflect the political ideology of the time that every citizen should contribute equally to the cost of the services which they consume. (There were some exceptions, for example students and those on very low incomes.) The system of local rates was well established and could, some argue, trace its beginnings back to the Poor Law Relief Act of 1601, although it can also be claimed that the modern system of local self-government was not established until the Municipal Corporation Act of 1835,[6] which empowered local authorities to levy local taxes. Both these claims however point towards the rates system as was, having a long and established, history.

In the 1974 election campaign, given the feelings of inequity with regard to the tax and the large variation in the basis for the tax across the country, the Conservative party pledged to replace the rates system. The election was however won by the Labour party, but the idea of change had taken root. After a number of events occurred which encouraged the search for a review of the rating system, a report proposing reform was published by the Department of the

Environment in 1986.[7] The three key parts of the proposal were that the system of government funds for local authorities would be revised, the local non-residential (business) tax would be replaced by a national business property tax and the residential property tax would be replaced by a flat rate community charge payable by all adults living in the relevant area. In 1989 the proposals were implemented almost unchanged in Scotland, by the now Conservative government elected in 1979, before being made law in England and Wales in 1990.[5]

13.3.2 The Project Issue

A review by the BBC[7] of Cabinet papers from 1989 and 1990 relating to the implementation of the Poll Tax and held at the National Archives, reveals that the tax was more complicated to administer than originally thought and proved to be immensely unpopular with the citizens of the UK who considered the tax to be unfair. This led to accusations that the Prime Minister was uncaring and acting as a virtual dictator 'riding roughshod over anyone who opposes you.'[7] Feelings against the new tax were far-reaching and lead to wide-spread protests, civil unrest, and extensive non-payment of the poll tax (only 90% of the tax due was collected by the end of the first year, while the Audit Commission estimated in 1991 that 21% of taxpayers required a court summons before paying the tax[5]). Administrative costs of collecting local taxes rose three-fold as a result. This cost increase, aside from the reluctance to pay, was linked largely to the change in the way in which the tax was calculated. Previously the rateable value of a property was calculated by way of a well established formula linked to the property itself. Unlike property, the registration of residents was more difficult as it depended on a variable parameter – the location of the declared residence of each taxpayer. Unlike property, residents will inevitably move around, the Audit Commission (1991) found that the percentage change in entries on the poll tax register of residents was 36% in rural areas and as high as 55% in areas such as inner London, thereby creating a much higher administrative burden.

Amongst the numerous issues with the poll tax, possibly the key one, contributing to the reluctance to pay the tax, was that the policy went ahead without the support or buy in of the electorate. Some analysis was done as to the number of winners and losers in terms of the amount of tax that each household would have to pay, but this analysis was superficial.[5] At the time of the proposal's publication the changes were not widely debated or understood and additionally they did not form a significant part of the 1987 general election campaign. Post the 1987 election, there had been plans to take a phased approach to the implementation of the new tax regime, but this approach to the project was abandoned when it became clear that a high proportion of those who would gain lived in the Conservative heartlands, and when the problems of running two tax systems alongside each other were clearly defined. In short it appeared simpler, and more beneficial in terms of Conservative voter reaction and the overall workability of the tax system, to launch the transformation in one go.

The new tax was not well received as evidenced by the non-payment rate detailed above, the impact of the new tax was also felt keenly within the Conservative party. The introduction of the tax within England coincided with a sharp decline in support for the Conservative party shown in contemporary opinion polls, and it has been cited as a major reason behind a subsequent challenge to the leadership of Margaret Thatcher who went on to be replaced by John Major in 1990.[5] Major was Prime Minister when the decision was later taken to abandon the poll tax in 1991 and his government gave the reason for this decision as being the "uncollectable" nature of the tax. The issues linked with administrating the tax were insuperable, and the lack of consideration of the way in which the tax would be collected in the project's early design stage proved to be a major flaw in the project's delivery.

13.3.3 How Could a Developed Business Readiness Process Have Helped?

One of the assumptions driving the change in tax policy was that the direct accountability between each adult paying the tax and the spending of the local councils, which was though by central government to be excessive, would drive increased allocative efficiencies. This assumption was untested pre-launch even though these tests could have been run. The phasing in of the new system and the running of pilot tests would have been wise as one of the architects of the original idea, William Waldegrave, the Conservative MP for Bristol, noted that once presented to the party leaders, the plan was introduced 'overnight in one go, which was never my plan and I thought they must know what there were doing - but they didn't.[8]' This suggests that the change was implemented before the readiness of either the system or the public to accept the new tax was known – in other words neither was ready and 'business readiness' had been far from assured.

Predictions of the financial implications were 'superficial'. If the testing had been more thorough it might have been possible to establish the truths relating to the predicted benefits of making local council spending more accountable to local residents and if established these arguments could have been used to sell the changes to the taxpayers and thereby persuading the public to accept the tax in greater numbers.

There were other factors which also contributed. Key actions such as early discussion of the tax changes within a wider environment had not been considered. Neither pilot tests nor a phased implementation had been fully thought through as a means of risk mitigation, or as a means of increasing the suitability of delivery plans. There had been an initial debunking of the idea of running a phased implementation but it was perhaps too quickly dismissed rather than considering a different approach to phasing the tax in. Without the use of these approaches, the understanding of the environment into which the project was to be delivered was compromised. More could have been known about the environment, more could have been known about the impact on the taxpayers and the likely reaction to the implementation, alterations to the approach could have been made before deployment, which could have helped ensure readiness and hence acceptance.

13.3.4 Conclusions on the Poll Tax Implementation Project

The implementation of the poll tax was a disaster. The government didn't take account of, nor appear to know, the likely reaction that the public and the strength of feeling that they were about to be faced with. The reaction to the tax and the levels of opposition it engendered made, in Major's words, the tax uncollectable and contributed significantly to its political downfall. The government's failure to prepare the public for the tax change and to ensure that the system was workable led to the London poll tax riots of March 1990, where up to 3,000 demonstrators clashed with police leaving 113 people injured of which 45 were police.[8]

Faced with the practical difficulties and the political pressures surrounding the introduction of the poll tax, Mrs Thatcher's political career was foundering. In November Michael Heseltine, an outspoken critic of the poll tax, triggered a leadership contest.[9]

Later that month, Prime Minister Margaret Thatcher stood down as leader of the Conservative party. She was replaced by John Major as Prime Minister of the United Kingdom on 28 November 1990 who went on to scrap the poll tax in favour of the council tax that remains in place today.[7]

Michael Portillo, who was at the time the Conservative local government minister, commented on the lessons learnt from the debacle and considered that a lack of 'thinking it through' was at fault. He was quoted[9] as saying that 'the lesson ought to be, think carefully before you do things' which is good advice that should be followed by all project managers considering deploying a new product and considering the vital issue of 'business readiness'.

13.4 Business Readiness and Project Closure Before the Birth of Modern Project Management*

(* In this section, the research relating to the construction of HMS Barham 1912–1915 mentioned in the introduction is drawn upon to help illustrate how the subject of this chapter has developed.)

When reviewing the 1912–1915 construction of HMS Barham, there are certain characteristics of the project which need to be considered. The ship was built by John Brown and Co. on the Clyde under the watchful eye of an inspector from the Admiralty along with detailed correspondence from the Admiralty. It was one of five ships built within the 1912 Elizabeth Class and as such the Royal Navy had specific practice in terms of accepting this Class of ships into service. The first of Class HMS Queen Elizabeth was commissioned on 22 December 1914, while Barham, the fourth of Class was commissioned some ten months later on the 19 October 1915, so any issues relating to the vessel's, or the accepting body's readiness to receive the Class into service would have been addressed by the time HMS Barham was commissioned. Taking a wider view of the issue, the commissioning of new ships into service is something that the Royal Navy had obviously been doing for many years preceding HMS Barham and so protocols would have been well established for this when it came to the commissioning of HMS Barham in 1915.

The issue of project closure was directly reflected in the records of John Brown. Project closure was not however signified by the document shown above in Figure 6.12. Figure 6.12 provides an extract from the John Brown and Co. Shipyard Diary giving details of a luncheon to mark the launch of HMS Barham on 31/12/14. The launch of a ship however is not the same as the commissioning of the ship. The launch of the hull, as recorded in the 31/12/14 record, is a key stage along the lifecycle. Further fit out is however still required. The John Brown and Co. Shipyard Diary provides detail of elements of this too, showing for instance the arrival of the gun mountings on 24–28/7/15 aboard the SS Aydon as part of the superstructure fit out.

HMS Barham did however go through a formal sign-off procedure and records from John Brown and Co. show that the Admiralty signed off a form agreeing the particulars of the Barham 'as built' on the 28th September 1915. The document is signed by both the dockyard and the Admiralty indicating that both sides agreed that the particulars of the ship were as required. The document itself is in effect, a sign-off/acceptance certificate for the project, as indicated by the document's heading, which reads 'this form is to be rendered on the completion of shipbuilding'.

The completed sign-off comes just eighteen days after the final issues relating to the gunnery tests findings were resolved, and in advance of HMS Barham being commissioned by the Royal Navy. The sign-off marks the end of the construction phase and the beginning of the transfer to the 'in-service' part of the lifecycle. As the project had a 'project lifecycle as opposed to an 'extended lifecycle, (see Chapter 4 for discussion of lifecycles), this marked the point of project completion for the Shipyard.

References

1 APM. (2019). *APM body of knowledge* (7th ed.). Association for Project Management.

2 Project Management Institute. (2021). The standard for project management and a guide to the project management body of knowledge (7th ed.). Project Management Institute, Inc.

3 Ohara, S. (2017). *P2M – A guidebook of project & program management for enterprise innovation* (3rd ed.). Project Management Association of Japan (PMAJ).

4 Calleam. (2024) *Why do projects fail? UK Government – The poll tax.* https://calleam.com/WTPF/?p=5515 [Accessed 10 January 2025].

5 Smith, P. (1991). Lessons from the British poll tax disaster. *National Tax Journal*, *44*(4), 421–436.

6 Wollmann, H. (2000). Local government systems: From historic divergence towards convergence? Great Britain, France, and Germany as comparative cases in point. *Environment and Planning C: Government and Policy*, *18*(1), 33–55. https://doi.org/10.1068/c9867

7 Margaret Thatcher Foundation. (2025). Green paper on local government finance ("Paying for Local Government") [origins of Community Charge]. https://www.margaretthatcher.org/document/217069 [Accessed 10 January 2025].

8 The BBC. (2015). *Poll tax was 'a serious mistake' - Waldegrave.* https://www.bbc.co.uk/news/uk-england-bristol-33594395. [Accessed 10 January 2025].

9 The BBC. (2016). *National archives: Thatcher's poll tax miscalculation.* https://www.bbc.co.uk/news/uk-38382416 [Accessed 10 January 2025].

Chapter 14

Project Failure and Complexity

Many projects suffer failure, and as a consequence, the project manager should always be on the look out for possible signs of failure and be ready to take appropriate action. Of course, there are some projects which seem set up to fail, or more accurately, not set up to succeed. Complexity can be a cause of project failure when it is not recognised and hence not managed proactively. This chapter considers a variety of approaches to recognising these issues within projects as early as possible in their lifecycle and reviews the approaches that could be taken to help avoid failure.

14.1 Definitions of Project Failure and Complexity

Project failure is not included in the list of definitions given in the APM's BoK7[1]. The PMI[2] link project failure with poor requirements management but it does not directly define project failure either. P2M[3] does discuss project failure and states that if, in the case of an internal project, operations post-delivery are entirely the same as before the project was launched then the project can be criticised as a failure. P2M also recommends that the project scope should include a quantitative measure to determine either success or failure. Project failure therefore reflects the project's delivery of the outputs defined in the project scope documents and is judged in terms of the prescribed baselined time, cost and quality deliverables (see discussion of the 'Iron Triangle' in Chapter 3).

The APM[1] state that the complexity of a project is a factor of the amount of interaction between different parts of the project and reflects the amount of uncertainty, the degree of interaction between stakeholders and the novelty involved in the project. P2M[3] add changes to customer needs, the emergence of rivals with new technologies which could not have been predicted, together with changes in supply markets and rare economic or political fluctuations potential sources of project complexity.

The PMI[2] assert that project complexity can be caused by a number of factors such as human behaviours, system interactions or uncertainty and ambiguity within a project which results in multiple interconnected influences that behave and interact in diverse and potentially unpredictable ways. It can be introduced into a project at any time by issues impacting multiple areas such as project scope, communications, stakeholders, risk and the emergence of new technologies. The PMI stress the need for project teams to be alert to any complexity, whether initial or emergent, within their projects so that approaches to the project can be modified to cope with the complexity whilst still maintaining the baselined deliverables outlined in the project business case. Overall definitions of project complexity can be many and varied meaning that there are a number of different approaches put forward for dealing with it within the delivery of a project.

DOI: 10.1201/9781003470748-15

14.2 How Is Project Complexity, Managed Within a Project?

14.2.1 Potential Bases of Project Complexity

Complexity can be drawn from numerous different aspects of a project. It can be derived from the way in which a project is managed or from a project's physical features. It can also be a reflection of how difficult the project objectives are to achieve or the extent of project risks and uncertainties.

The Global Alliance for Project Performance Standard (GAPPS) identify seven causes of project management complexity within their Framework for Performance Based Competency Standards[4] for project managers, these are listed as:

- The stability of the overall project context particularly the stability of the project lifecycle, the stakeholder engagement, the methods and approaches used together with the wider environment into which the project is delivered.
- The number of different disciplines, methods, or approaches used in delivering the project, which could be seen as a measure of the size of the project.
- The number and size of project interfaces giving rise to legal, social, or environmental implications, also potentially a measure of project size.
- The number of interfaces between the project and other organisations or disciplines can also add to complexity just as those listed in the point above will.
- The overall expected financial impact, either positive or negative, on the project's stakeholders. The more important the project is in these terms, the more scrutiny is likely to be on the project, which in turn will generally add to the level of complexity.
- The higher the level of strategic importance of the project to the organisation, the higher the number of interfaces it is likely to have and the higher the impact of those interfaces will be for the organisation.
- The higher the number of stakeholder is, then the higher the chance that there will be divergent views in terms of the expected outcomes. This will in turn, increase the degree of complexity in terms of delivering a successful project.

Other sources cite additional causes of project complexity, such as a project's physical features. These could easily impact the complexity of the project if, for example, the project becomes short of time, then the activities might be re-sequenced meaning that the overall degree of project difficulty and or complexity could increase. The causes of project complexity are not fully agreed upon within the project management literature and the following list, together with that above, although far from exhaustive, notes some additional areas of potential complexity which the project manager would be wise to consider:

- Uncertainty or degree of unusualness associated with the project goals.
- Ambiguity or rapid change in project deliverables or delivery systems.
- The level of technical challenges within the project.
- The number of variables or unknowns linked with the end state into which the project delivers.
- The number and variety of end states in which the project deliverables will need to be used.

14.2.2 Signs of a Failing Project

The lists in the previous section represent some potential sources of project complexity. It is not always complexity though that causes a project to fail. There are a number of signs which can warn the project manager that a project is in danger of failing. When a project begins to overrun its budgets, when there is significant slippage in the schedule and when quality problems mean that the project outputs do not meet the client's needs then the project is likely to be in danger of failing. Staff churn may also increase as staff become desperate to leave a sinking ship! In addition to the challenges created by project complexity, these symptoms could be caused by one-off trigger events or by behavioural causes. A trigger event is where individual actions or mistakes occur and impact the project's chance of success. Behavioural causes on the other hand relate to patterns of behaviour or structural issues within the project and its management, these are unlikely to be one-off events. Often though, what allows these causes of project management to occur are issues that can be collected under the heading of dysfunctional or ineffective decision making, such as poor situational awareness, inexperience, lack of training, political issues or misaligned goals and poor communications.

The National Audit Office and the Office of Government Commerce agreed an interesting list of project failures causes. Interestingly, nowhere in the list is complexity mentioned. The full list of causes focuses on: the lack of a link between the project and the organisation's strategy, the lack of clear senior management ownership of the project, the lack of effective stakeholder engagement, the lack of project management skills particularly within the management of risk, too little attention to breaking implementation into manageable steps, a lack of focus on long-term value for money, a lack of both understanding and contact with others in the supply chain, and a lack of project team integration between client, project team and the supply chain.

This looks to be a daunting list, but the key point must be that each component of it is a potential cause of project failure because something is lacking. The list above should therefore become a useful check list to help the project manager ensure the success of their project.

14.2.3 Ways to Address Project Complexity

There are approaches that can be taken to help minimise the problems caused by project complexity. Some of these begin with the processes and techniques used to address areas such as risks. If risks are mitigated effectively (see Chapter 9,) then some areas of complexity at least, may not develop.

There are other possible sources of complexity within projects which can be minimised by giving them consideration at the time when the project is established and by reflecting them in the way in which projects are established, thereby minimising the potential impacts of complexity on project outcomes. Two of the key areas to consider are the project lifecycle used and the style and approach chosen to manage the project.

14.2.3.1 Project Lifecycle Considerations

When the project lifecycle is considered thought should be given to the level of new technology involved in the project, and how established and well known by the market the project output is and hence the degree of complexity which could be generated. The use of new technology, or a high level of technology within project outputs, will inevitably add unknown factors to the profile of a project, meaning that a more iterative approach to the life cycle is likely to be appropriate. A more iterative approach allows for an increased number of design cycles within the project lifecycle which in turn, will enable more testing and greater certainty in terms of the new technology's implications. If the environment into which the project outputs are delivered is relatively unknown, then an iterative approach will also allow for the design to be frozen at a point whilst certainty about the delivery environment evolves, or is tested.

Multiple iterations allow for the identification of what worked well in each iteration and equally what did not work well, enabling a refining of the project deliverables. Key stakeholders (Chapter 6) should be included in discussions relating to the results of each iteration, this not only has the potential to help clarify future project direction, but also to test any assumptions made to date thereby sharpening the view of both the project end deliverables and the environment in which they are delivered.

If a waterfall project lifecycle were to be used rather than an iterative approach, then the complexity and surrounding uncertainties of the new technology and environment would be likely to necessitate numerous project re-sets as the implications of the technology and environment became apparent. In the worst case this mis-selection of the lifecycle approach could lead to project failure. Recognising the exploratory nature of the project and proceeding little by little, in an iterative manner, will enable the project team to focus on a specific relatively known part of the project output thereby reducing the level of complexity being dealt with at any one time and increasing the chances of a successful outcome.

14.2.3.2 Management Style Considerations

There is a perhaps a tendency these days to lean towards the use of an informal project management style. This however can be detrimental where the project outputs are delivered into a wide range of geographic areas or into a number of different organisations. A project could for example be delivering a new product into several European countries governed by various local laws, multiple languages and numerous local customs. In this scenario, the project manager would need to consider how best to add certainty and clarity to the project given the level of complexity which these differences could lead to. The greater the geographic area a project is delivered to, or the greater the range of users for the project's deliverables then the larger the project team is likely to be and hence the more challenging, or complex, it will be to manage the team in an informal manner. A formal style of project leadership will be required to ensure that all parts of the project team have the same view of what is required from them. Rather than relying on informal walk arounds and chats with team members about an evolving project design, the development of such a project means that directives and decisions should be formally documented and made available to the whole team in a central directory. It requires an approach which may appear less flexible, but it will add significant clarity and certainty of project requirements. The higher the level of project complexity, then the more formal the management style is likely to be.

The opposite to a large, formalised project type is perhaps a small project where the complicating factor is the speed of deployment required. Faced with this type of project and the complexities that this brings, consideration should also be given to the make-up of the team and the degree of bureaucracy involved. A project tasked with delivering at a fast pace will need to be autonomous and ideally to include top management in the project team to enable decisions to be taken quickly, without the sometimes time consuming need to refer decisions upwards. It can be seen that this type of project should be set up in order to minimise the levels of bureaucracy and to facilitate quick decisions, which will be enabled by making the correct choices about management structures at the outset of the project.

14.2.4 Does the Project Team Selected Have the Required Competencies?

Managing complexity does not just depend on the processes chosen. It will of course also be impacted by the members of the project team themselves and the level of project experience that they have.

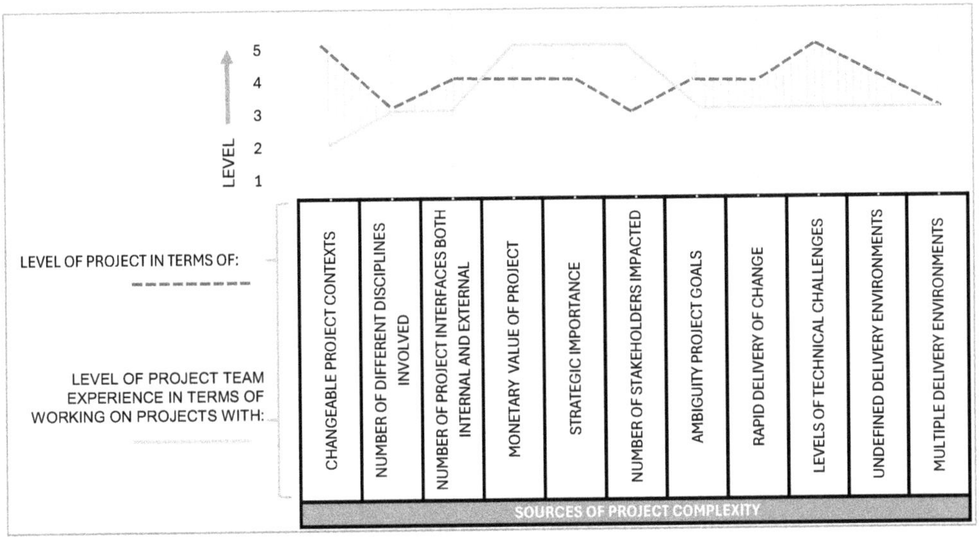

Figure 14.1 Project complexity/team experience graph.

Figure 14.1 above shows a diagram plotting the complexity profile of a fictitious project against the experience of the project team delivering it.

The graph in Figure 14.1 takes the potential sources of project complexity discussed in Section 14.2.1 and places them at the bottom of the Figure. The nature of the project is then reviewed and the appropriate ranking from 1 to 5 is given for each component and plotted using the black dashed line. The experience of the project team members as a whole, is then also reviewed and plotted using the solid grey line.

Where the experience of the team is below the profile required by the project, the graph has been shaded. This highlights the potential areas of difficulty that the project may encounter as a result of the project team not having had prior exposure to, or sufficient experience of, these sources of complexity.

Having identified the gaps, there are a few alternative measures that could be taken dependant on the seriousness of the shortcomings identified. The most extreme would be to seek the replacement of some project staff by others with greater relevant experience. Alternatively, more experienced staff could be added in as extra staff rather than as replacements, or existing staff could be offered training in these areas. The least that should be done, assuming none of the above are instigated, is that the gap identified should be entered into the risk log (see Chapter 9) as a potential source of project risk.

14.3 Project Review: The Impact of Complexity on the Project: The Great West Electrification Project (GWEP)

14.3.1 The Project Background

The Great West Electrification Project (GWEP), approved in 2013 and originally priced at £874m,[5] was a project aimed to improve services along the rail route from London to the west and south-west of England and into South Wales,[6] thereby delivering on the UK Department for Transport's commitment to increase the amount of the rail network covered by 25 kV Overhead Line Electrification installations. Faced with a continued rise in the expected usage for a service which was already amongst the most crowded in the country, the need to increase capacity and provide fast and reliable journeys was squarely behind the project's instigation.

The project focused not just on delivering the rail electrification, but also included multiple other elements, such as the delivery of new electric trains, the construction of 50 new auto-transformer sites, the updating of the signalling system along the route, the structured clearance of around 160 bridges and four tunnels together with the wiring of over 1,000 km of track[7] on the Great Western Mainline route.

The full extent of the route ran from Maidenhead, west of London, to Cardiff in South Wales and interfaced with the Crossrail project at Maidenhead. The project started in March 2014 and the Bristol, Newbury and Oxford lines were targeted to finish by 2016, followed a year later by the line to Cardiff.

The key benefits of the project were identified[7] as the provision of an extra 20% capacity, reduced track damage due to trains that were lighter than the previous diesel trains, and the trains themselves would also be quieter and cheaper to run with less maintenance required and lower energy costs.

The GWEP project itself was part of a programme, or a group of related projects – the Great Western Route Modernisation programme (GWRM). Alongside the GWEP project, the main elements of the programme included:

- The Western Capacity Enhancement Programme (WCEP) involving infrastructure works aimed at providing additional fleets of trains
- The Intercity Express Programme (IEP) aimed at providing the Great Western Main Line with additional rolling stock.[9]

There were also interfaces with other programmes such as Crossrail[9] due to the large number of 'interactions between Crossrail and the Great Western franchised London commuter operations.'

14.3.2 The Project Issue

The GWEP project ran into a number of difficulties causing delays to its completion and prompted cost estimates to more than triple from £874m in 2013 to £2.8bn in 2018.[8] The increasing level of costs put pressure on the project and caused the scope to be reduced. Electrification of parts of the network were put on hold, whilst some elements such as the electrification between Cardiff and Swansea were cancelled completely.[8]

There were of course many aspects of the project which caused the increases in cost and time. Some were linked to a failure of management to identify and to reflect the complexity of the project. This was brought about in part as a result of it being a component of the larger GWRM programme which led to an initial failure to plan and manage all the component projects in a sufficiently joined up way. The complexity of the project was also impacted by the requirement to carry out the complicated infrastructure works not only on a live railway, still transporting passengers and cargo, but also passing through areas of outstanding natural beauty and involving listed buildings in heritage areas.[6]

Another layer of complexity weas added by multiple parties making key decisions. The Department for Transport and Network Rail entered into a contract to purchase trains for the Intercity Express Service and in doing so, it created fixed deadlines for the electrification of the network. This would have impacted the GWEP project even if the approach to the electrification of the line had been fully understood, but Network Rail had only just recognised that a revised approach to the electrification of the network would need to be developed. This lack of complete understanding of the project interfaces meant that project schedule was not informed by a bottom up appreciation of the tasks involved, which directly impacted the accuracy and deliverability of the schedule there by adding additional time pressure to the project.[6]

A review of the NAO report on the modernisation of the Great Western Main Line by the Welsh Parliament,[10] also highlighted an 'unusually complex portfolio of schemes, which were subject to poor initial scope definition and ongoing scope creep' which in turn contributed to 'costing errors, unanticipated interdependencies, (and) a lack of consideration given to deliverability'. Part of this added complexity was triggered by the Network Rail's underestimation of how many bridges would need to be rebuilt or modified and the need to obtain planning permissions and other consents for works relating to protected species or listed buildings. More than 1,800 separate consents were required for such works.[6]

14.3.3 How Could an Improved Approach to Managing Complexity Have Helped the GWEP Project?

Section 14.2.1 lists the seven causes of project management complexity identified by GAPPS. The review of reports on the project noted in Section 14.3.2, that there were three causes of complexity which stood out and which were closely related. Firstly, the number and size of interfaces that the project had was noted as a potential source of legal, social and environmental implications, whilst the number of interfaces directly between the project team and other organisations or disciplines was noted as a second cause of complexity. The third source of complexity derived from the key strategic importance of both the GWEP project and the larger programme of which it was part. The higher the level of strategic importance of the project to an organisation, the higher the number of interfaces it is likely to have both inside and outside the project, and the greater the impact of those interfaces will be for the organisation. The question must be asked why these interfaces were not identified at an early stage? The answer is perhaps in observations detailing how prior to 2015, the project was not planed nor managed 'in a sufficiently joined up way' in relation to the other projects which made up the GWRM programme.[6] Had it been, then there would have been a greater understanding of the complexity and hand-offs necessary between areas of the project(s) for it to succeed.

The level of technical challenges within the project was initially understated. The electrification of the Severn Tunnel was one such element. The impact of humid and salty conditions within the tunnel caused significant problems due to dirt and pollution from the trains passing through the tunnel sticking to sensitive electrical equipment,[8] the extent of which was initially under appreciated. The deterioration of electrical components meant that power had to be switched off, and the issue's initial lack of recognition resulted in Network Rail spending 3 years and £20m employing external rail electrification experts in order for the tunnel to resume electric operation. This again points to a lack of time initially taken to fully understand the project and its complexities, reflected in the poor planning at the time and late development of a business case.[6] Perhaps the high number of project stakeholders and the resulting increased possibility of divergent views also added to the complexity and complicated planning. There is nothing substantial written in the reports quoted that reflects how well the communication plan was executed between the stakeholders. With the late development of an overall business case, it can be reasonably assumed that in the early stages this was not a key focus of the project, potentially resulting in additional complexity caused by multiple stakeholders seeking to influence the project in multiple ways and directions.

Overall, more time spent in up planning up front would have produced a greater understanding of the project and allowed active consideration of the potential sources of complexity which would have made the project easier to deliver. The NAO's report[6] says that Network Rail didn't develop a critical path for the project before starting to deliver it, nor did they 'conduct sufficiently detailed surveys'. Network Rail were adjudged to have failed in their approach to the technical challenges of the project and the risks of using new technology. If these could not have been identified at the outset of the project, then additional staff used to working with similar technical challenges should have been brought in and more testing cycles employed within a more iterative project lifecycle.

14.3.4 Conclusions on the GWEP Project

The Welsh Parliament[10] highlighted the poor initial scoping of the project, the unanticipated interdependencies, scope creep and poor project governance within the project. All of these issues point to a lack of understanding of the project complexities and hence a lack of focus on resolving or tailoring the project to best deal with these complexities, indeed it is reported[11] that the then heads of the NAO asserted that whilst the modernisation of the route had potential to deliver benefits, the poor management of the projects meant that these benefits were delayed. If a benefit is delayed, then it will generally be reduced in size as it is a function of the capability delivered and the length of time it is in operation, and so the gap between the costs and benefits in the business case will begin to be eroded. Failure to understand and manage the project appropriately with the right processes and expertise therefore led the project to underdeliver in terms of the overall level of benefits that were available.

14.4 Project Complexity Before the Birth of Modern Project Management*

(* In this section, the research relating to the construction of HMS Barham 1912–1915 mentioned in the introduction is drawn upon to help illustrate how the subject of this chapter has developed.)

The Dreadnoughts, built as part of the WW1 Grand Fleet, were at the cutting edge of design, so much so that it was said that they made all older ships obsolete.[12]

This was due in part to their ability to out-gun all previous naval vessels and their fire power continued to develop as further Dreadnoughts were laid down during the period. The vessels were therefore complex undertakings, not just due to the advanced capability for the time, but also due to the monetary value of the project, the strategic importance and the rapid speed of delivery which was required during a time of war. (See Figure 14.1 for a list of potential factors impacting project complexity.)

Whilst complexity is not a discussion found in the archives linked to the Case Study of HMS Barham, there were a number of approaches taken in the construction of the Dreadnoughts which would have helped to address the inherent complexity characteristics of the project. Firstly, each year in the early part of the twentieth century groups, or classes of ships were built, normally of 4 or 5 closely related vessels. HMS Barham was part of the 1912 programme[12] – the Queen Elizabeth Class consisting of HMS Queen Elizabeth herself, and HMS Warspite, Valiant, Barham and Malaya. Because construction was staggered, with start dates a few months between each other, there was the opportunity to learn from earlier iterations of the class, and to refine designs, addressing any areas of complexity. The first of class was built in HM Dockyards which would enable the initial stages of the concept to be proved under the watchful eye of the

Admiralty and the Admiralty inspectors before issuing designs to the commercial yards, (HMS Queen Elizabeth was built at Portsmouth, HMS Warspite was built at Devonport respectively, HMS Valiant was built by Fairfield Shipbuilding & Engineering Co, HMS Barham was built at the Clydeside yard of John Brown and HMS Malaya was built by Armstrong Whitworth & Co.)

The vessels shared many common parts which meant that understanding of their interoperability would grow as the class was delivered. Perhaps though, the greatest mitigation against potential problems linked to complexity would have been the experience of those involved in the design and construction. Figure 14.1 above considers the profile of the team in terms of their experience of similar projects and identifies the gaps between that experience and the profile of the project as possible sources of risk. The initial design of the class was drawn up under the supervision of Sir Philip Watts, Director of Naval Construction (1902–1912), and the yards where the ships were constructed had been involved in naval construction for many years. As a result, both the companies and staff involved in the construction brought a high level of skill and experience to the project meaning that they would have been better equipped than most to deal with the levels of complexity involved.

One issue however did cause problems for the Dreadnoughts at the Battle of Jutland. The 15-inch guns in ships like HMS Barham were capable of firing shells further than any previous ships. The ability to accurately aim them however was less well developed and the shells did not hit their targets as often as required. The command was therefore given by Beatty to fire as many shells as quickly as possible.[13] Unfortunately, in order to do this and to speed delivery of shells from the magazines to the guns, the doors to the armoury were propped open. This did increase the rate of fire but with disastrous consequences. Two ships, the Queen Mary and the Invincible are recorded as having suffered explosions in their magazines,[12] after having received direct hits to their gun turrets. The exploding German shells resulted in flash fires, down through the armoury doors, causing the magazines to explode. Investigations post the battle of Jutland put the 'blame on ship's staff for having too many exposed charges and generally slack procedures, encouraged by senior officers to whom rate of fire was everything.'[12] The propping open of the doors to the magazine was one of these 'slack procedures', the purpose of which was to speed the delivery of the shells from the magazines to the guns.

This highlights that the issues of interoperability between the firing and aiming of the guns had not been sufficiently understood, and as detailed in Section 14.2.1 above is an example of the 'number and size of project interfaces' being a measure of project complexity. There are of course several other sources of complexity that can be drawn from the list at Figure 14.1. The monetary value of the project was high – the budget had to be approved by parliament as part of the Naval Estimates for the 1912 programme. The strategic importance was clear with the looming First World War, and of course the technology used was, at the time cutting edge. All these aspects added to the complexity of the project and would have needed to be fully understood and managed well if, as it was, the ship was to be delivered to the Admiralty on time.

References

1 APM. (2019). *APM body of knowledge* (7th ed.). Association for Project Management.

2 Project Management Institute. (2021). The standard for project management and a guide to the project management body of knowledge (7th ed.). Project Management Institute, Inc.

3 Ohara, S. (2017). *P2M – A guidebook of project & program management for enterprise innovation* (3rd ed.). Project Management Association of Japan (PMAJ).

4 GAPPS. (2007). *A framework for performance based competency standards for global level 1 and 2 project managers.* https://www.projectmanagement.com/content/attachments/Primus1_201011023434.pdf. [Accessed 10 November 2024].
5 Topham, G. (2015, October 21). Cost of Great Western mainline electrification project triples to £2.8bn. *The Guardian.* https://www.theguardian.com/uk-news/2015/oct/21/cost-to-electrify-great-western-mainline-triples-to-28bn-risking-other-schemes [Accessed 18 November 2024].
6 The Comptroller and Auditor General. (2016, November). *Modernising the Great Western railway.* National Audit Office. https://www.railwaysarchive.co.uk/documents/NAO_GWElectrificationFull2016.pdf [Accessed 6 November 2024].
7 Network Rail. (2025). Great Western electrification. https://www.networkrailconsulting.com/our-capabilities/network-rail-projects/great-western-electrification/ [Accessed 8 November 2024].
8 International Railway Journal. (2020). *Network rail completes Great Western electrification.* https://www.railjournal.com/passenger/main-line/network-rail-completes-great-western-electrification/ [Accessed 10 November 2024].
9 Department for Transport. (2022). *Great western route modernisation first post-opening evaluation – Final report.* https://assets.publishing.service.gov.uk/media/5fc51d7b8fa8f54755392863/full-business-case-hs2-phase-one.pdf [Accessed 15 November 15 2024].
10 Welsh Parliament. (2016). *Great western electrification plans derailed? A national audit office report.* https://research.senedd.wales/research-articles/great-western-electrification-plans-derailed-a-national-audit-office-report/ [Accessed 15 November 15 2024].
11 Smale, K. (2016). *Damning report into Great Western electrification.* https://www.newcivilengineer.com/latest/damning-report-into-great-western-electrification-10-11-2016/ [Accessed 15 November 15 2024].
12 Brown, D.K. (2010). *The grand fleet: Warship design and development 1906–1922.* Pen and Sword.
13 Jellicoe, E.J.R.J. (1919). *The grand fleet, 1914–1916: Its creation, development and work.* Doran Company.

Glossary

AC	Actual Cost (EVM)
APM	Association for Project Management
BAC	Budget at Completion (EVM)
BoK	Body of Knowledge
CIOB	The Chartered Institute of Building
CPI	Cost Performance Index (EVM)
CSCR	Committee on Shuttle Criticality Review
CSR	Corporate Social Responsibility
CV	Cost Variance (EVM)
DoH	Department of Housing
DoT	Department of Transport
EAC	Estimated Cost at Completion (EVM)
ETC	Estimated Time at Completion (EVM)
EV	Earned Value (EVM)
EVM	Earned Value Management
FBB	Flughafen Berlin Brandenburg GmbH
FMC	Ford Motor Company
GAPPS	Global Alliance for Project Performance Standard
GWEO	The Great West Electrification Project
GWRM	Great Western Route Modernisation programme
HS2	High Speed 2
IoG	The Institute for Government
IRR	Internal Rate of Return
ISO	International Organization for Standardization
MSP	Member of the Scottish Parliament
MPM	Modern Project Management
NAO	National Audit Office
NPV	Net Present Value
P2M	Program & Project Management (written by the PMAJ.)
PM	Project Management / Project Manager
PMAJ	Project Management Association of Japan
PMBoK	Project Management Body of Knowledge (written by the PMI.)
PMI	Project Management Institute
PMP	Project Management Plan
Programme	a group of related projects

PV	Planned Value (EVM)
SPI	Schedule Performance Index (EVM)
SV	Schedule Variance (EVM)
TAC	Time at Completion
TCQ	Time Cost and Quality
TfL	Transport for London
WBS	Work Breakdown Structure

Index

Note: **Bold** page numbers refer to tables and *italic* page numbers refer to figures.

For Product Safety Concerns and Information please contact our EU representative GPSR@taylorandfrancis.com
Taylor & Francis Verlag GmbH, Kaufingerstraße 24, 80331 München, Germany

www.ingramcontent.com/pod-product-compliance
Ingram Content Group UK Ltd.
Pitfield, Milton Keynes, MK11 3LW, UK
UKHW052302060226
467784UK00022B/552
9781032739175